Project Management Methodologies, Governance and Success

Insight from Traditional and Transformative Research

Best Practices in Portfolio, Program, and Project Management

Series Editor
Ginger Levin

Project Management Methodologies, Governance and Success

Insight from Traditional and Transformative Research

Robert Joslin

CRC Press
Taylor & Francis Group
Boca Raton London New York

CRC Press is an imprint of the
Taylor & Francis Group, an **informa** business

AN AUERBACH BOOK

CRC Press
Taylor & Francis Group
6000 Broken Sound Parkway NW, Suite 300
Boca Raton, FL 33487-2742

First issued in paperback 2022

© 2019 by Taylor & Francis Group, LLC
CRC Press is an imprint of Taylor & Francis Group, an Informa business

No claim to original U.S. Government works

ISBN 13: 978-1-03-247567-7 (pbk)
ISBN 13: 978-1-4665-7771-8 (hbk)

DOI: 10.1201/9780429071416

Visit the Taylor & Francis Web site at
http://www.taylorandfrancis.com

and the CRC Press Web site at
http://www.crcpress.com

Contents

List of Figures

List of Tables

Foreword

One of the most encouraging signs of "health" in an academic or professional field is often the progress being made to identify and define the unique theories that underpin and help us understand it. This sense-making process is crucial for newcomers to the field as well as established scholars and researchers because it serves as a means for organizing our body of cumulative knowledge and ensures that work being pursued relates to a "particular cognitive problem" held in common (Cole 1983, p. 130; Söderlund 2012). One of the fascinating aspects of the project management discipline has been the myriad theoretical perspectives it has spawned, in a relatively short time, as scholarly interest in the field has grown. These perspectives are varied and offer critical insights into models for understanding projects, such as the Management of Projects (Morris 2013), temporary organizations (Lundin and Söderholm 1995), p-form organizations (Söderlund and Tell 2012), project ecologies (Grabher 2004), governance (Müller 2009), and so many more. These theories reflect the results of a fascinatingly complex setting—projects and temporary organizations—coupled with a host of challenges in supporting contexts, including leading temporary teams (behavioral), identifying and managing stakeholders (political), identifying new processes and methods for practice (technical), understanding what factors determine project success (integrative), and so on. As project management scholarship enters a more developed and mature phase, it is heartening to witness the variety of manners with which theorists have begun to investigate and better understand the discipline.

Corresponding to the rise in theory development in project management has been a concomitant need to establish a research tradition that embraces multiple perspectives, allows scholars to investigate phenomena from a variety of contexts, and is flexible enough to embrace the alternative ways in which scholars seek to make sense of the project management field. Recent work in the field is intended to start this discussion moving forward by introducing several important works seeking to establish a research tradition shaped by various theories of project-based work (Lundin and Hallgren 2014; Drouin et al. 2013; Pasian 2015). In contributing to this developing field, Robert Joslin's work, *Project Management Methodologies, Governance and Success*, provides a welcome and timely addition.

A critical challenge with books such as this is the need to establish both a theory and method for undertaking research in project settings. That is, some works have done a fine job of relating the current state of project theory but lack robust discussion of methods for researching these settings. Other books offer strong analysis and recommendations for pursuing effective research methods but lack the discussion of the specific and challenging context of project-based work settings. That is, they are excellent primers on research but seem to implicitly forget that research methods by themselves are not sufficient, absent a clear understanding of the limitations and opportunities provided within project organizations or to address project-specific questions. Written in an academic but highly accessible style, one of the major achievements of this present work has been Dr. Joslin's linking together in a cogent manner the diverse themes of research theory and design, projects and project success (the critical dependent variable), and organizational governance. In effect, this book demonstrates that to fully understand how to undertake research in projects, theory and method are inextricably interwoven.

As the book clearly notes, the goal of practicing project managers and scholars alike lies in solving the puzzle of how to manage projects toward successful completion. Alas, understanding what comprises project "success"—seemingly such an innocent question—has come to represent one of the thorniest problems we face. Who determines success? At what point in time is success best measured? How do diverse stakeholders define success? What happens when their perspectives collide? What is the difference between *project success* and *project management success*? These are surprisingly complex problems. For every principle or rule we posit, a brief investigation reveals that there are numerous exceptions, muddying the waters and making these ideas increasingly opaque. This book, *Project Management Methodologies, Governance and Success*, addresses this challenge head-on, putting into proper context the critical issues that shape our understanding of the project management research process. Employing an idea referred to as "philosophical triangulation," Dr. Joslin shows us how to overcome the weaknesses or intrinsic biases that disrupt and minimize the impact of so much organizational research. Thus, understanding organizational governance and success within their proper context permits scholars to identify the best methods for researching project-based work challenges.

I applaud the publication of this book, as both a genuine achievement in its own right and a further signal of the strength of the project management discipline. Our field continues to excite scholars, offer invaluable insights into organizational and behavioral interactions, and provide important interpretive evidence of the direction in which future commercial and economic vigor lies. With all signs pointing to an increased use of and interest in project-based work in modern organizations, the better scholars are able to make sense of the current state of the field through theory development and empirical investigation, the more successful projects promise to become. This book is a welcome addition to our field and will be, I am sure, an important work and source for future reference well into the future.

Jeffrey Pinto, PhD
Penn State University

References

Cole, S. (1983). The hierarchy of the sciences? *American Journal of Sociology,* 89: 111–139.

Drouin, N., Muller, R., & Sankaran, S. (2013). *Novel Approaches to Organizational Project Management Research.* Copenhagen: CBS Press.

Grabher, G. (2004). Temporary architectures of learning: Knowledge governance in project ecologies. *Organization Studies,* 25: 1491–1514.

Lundin, R., & Hallgren, M. (2014). *Advancing Research on Projects and Temporary Organizations.* Stockholm: Liber.

Lundin, R., & Söderholm, A. (1995). A theory of the temporary organization. *Scandinavian Journal of Management,* 11: 437–455.

Morris, P. W. G. (2013). *Reconstructing Project Management.* Chichester, UK: John Wiley and Sons.

Muller, R. (2009). *Project Governance.* London: Routledge.

Pasian, B. (2015). *Designs, Methods and Practices for Research of Project Management.* Surrey, UK: Gower.

Söderlund, J. (2012). Theoretical foundations of project management, in Morris, P., Pinto, J., and Söderlund, J. (Eds.). *The Oxford Handbook of Project Management.* Oxford, UK: The Oxford University Press, 37–64.

Söderlund, J., & Tell, F. (2009). The p-form organization and the dynamics of project competence: Project epochs in Asea/ABB, 1950–2000. *International Journal of Project Management,* 27: 101–112.

Preface

Project management methodology (PMM), practices, and guidelines are the only explicit information that project managers have and, when properly maintained, should reflect the most current knowledge and guidance to achieve repeatable, successful project outcomes. Despite more than 50 years of research in the field of project management, project success rates are persistently low, and when viewed at a macro level, the impact can be seen at a country level.

The aim of this research is to advance the understanding of PMMs through a new perspective on PMM by:

- Developing a natural- to-social-science comparative model
- Using proven research methods to determine whether there is a relationship between the PMM and project success that is influenced by project context, notably project governance
- Identifying whether there are similarities and differences in the observed phenomena from the comparative and the conventional mixed-method approaches, and if so, to explain why
- Determining the direct impact of project governance on PMM and project success

First, this study will increase the understanding of PMMs and the factors that impact the effectiveness of a PMM used to support a project in order to achieve repeatable, successful project outcomes. Second, this study will increase the understanding of project governance influence on the direct and indirect impact of a PMM's effectiveness and completeness, the PMM's influence on project success, and project governance's direct impact on project success.

The first study, (a conceptual prestudy) explored whether it was possible to develop a natural-science comparative to social science, notably project management. The result was a comparative model that was tested using theory building based on complex adaptive systems (CAS). The findings showed it is possible to create a comparative that can be used to identify new phenomena and explain existing phenomena based on a natural science perspective, which cannot be easily explained using traditional social science perspectives.

The second study used a theoretically derived research model to qualitatively investigate whether different project environments impact the relationship between a PMM's elements and project success, and whether this relationship is influenced by the project environment—notably project governance. The findings showed that there is a positive relationship between a PMM's elements and the characteristics of project success and that the influence of the project environment—notably project governance—does influence the effectiveness of this relationship.

The third study quantitatively determined the relationship between a PMM and project success and the influence of project governance on this relationship. The study found that the successful application of a comprehensive PMM, where the term *comprehensive* is taken to mean including or dealing with all or nearly all the elements or aspects of something, accounts for 22.3% of the variation in project success. Project governance has an indeterminable effect on the relationship of PMM and project success. However, project governance does have a direct influence on the establishment and evolution of a PMM and whether it is a comprehensive PMM or one that needs to be supplemented by the project manager.

The fourth and final study explores the role of project governance on project success. The findings show that (1) a stakeholder-oriented project governance accounts for 6.3% of the variation in project success, and (2) project governance structures that are more control behavior–outcome orientation have no impact on project success.

Chapters 1 through 4 set the stage for the research. Chapters 5 through 9 present the research. The two concluding chapters discuss how the research findings further theory, as well as the practical implications of the research findings.

Throughout the book, the term *methodology,* when used in the context of projects, has been abbreviated to PMM, meaning project management methodology.

Robert Joslin
Wollerau, Switzerland

Acknowledgments

Thanks to my wonderful supportive and loving wife, Thanya, who has put up with my long hours of research, writing, and reviews. To my children, Ethan and Sydney, for tactfully changing the subject from research to something a lot more interesting.

To Ralf, my supervisor and friend, who has guided and supported my journey of discovery. To SBS Swiss Business School and especially Dean Dr Bert Wolfs, who is a visionary in so many ways, seeing and knowing how to get the best out of everyone.

And finally, to the special people around the world who have been supportive in this long and challenging journey (Alberto, Tea, Khoren, Lilly, Darko, and the rest of the team).

About the Author

Robert Joslin PhD, IPMO-M, IPMO-E, PMP, PgMP, PfMP, CEng, MIEEE, MBCS, is a professor, head of the research advisory board at SBS Swiss Business School, and the founder of AIPMO (Associational of International Project Management Officers).

Robert has considerable experience in designing, initiating, and project/program management delivery of large-scale business transformation, reengineering, infrastructure, and strategy development, having received several awards for ideas and product innovation. Robert presents at conferences as a keynote speaker and publishes books, book chapters, and research papers, winning the Best Research Paper Award at EURAM 2017.

Robert is in the process of architecting and co-authoring AIPMO's Body of Knowledge, which comprises several books in the area of PMO and related topics. Previously, he has been a consultant in a wide range of industries including telecom, banking, insurance, manufacturing, and government while working for McKinsey & Co, Logica, and his own consulting and product development companies.

Robert lives in Switzerland with his wife, Thanya, and two children, Ethan and Sydney.

Chapter 1

Introduction

This chapter introduces the need for new research methods and provides new insights into how project management methodologies (PMMs) may be better selected and applied to improve the chances for project success.

The chapter then describes the aims and objectives of the two parts of the research: the prestudy (natural-science comparative) and the main study—the impact of a PMM on project success with the determination of whether project governance impacts this relationship. The main study also investigates the impact of project governance directly on a PMM and then directly on project success. After this, the research focus is described and concludes with a summary of the research papers.

1.1 Background and Research Context

This section provides a background on PMM and its influence on project success.

1.1.1 Need for New Research Methods

The methods and techniques used today in project management research provide well-established frameworks for designing and executing research studies. The availability of methods and the acceptance of research paradigms mold the design of research studies and, through this, create a discipline. However, the success of these established approaches have some unforeseen consequences in terms of constrained academic thinking. The questions asked are often limited by the methodological starting positions and possibilities (Williams and Vogt 2011). The nature of a research design impacts research results, and the repetitive use of similar designs leads to almost predictable results. These constraints can be seen in many of the papers being submitted to academic journals but, more importantly, restrict reviewers in the peer-review process during which papers that demonstrate fresh and innovative thinking are rejected. Contemporary methods which have been developed and applied in many fields of scientific activities have provided for the development of new theories that challenge established theories and provide for fresh and alternative explanations of phenomena (e.g., Alvesson and Sköldberg 2009; Flyvbjerg 2001).

One area in project management research that would benefit from an alternative perspective using contemporary methods is that of addressing persistently high project failure rates. This would be especially beneficial for projects that are using PMMs that are also suffering from high project failure rates (Wells 2012). Project failure rates and PMMs are described in more detail in Section 1.1.2.

A prestudy was carried out to create a natural-science comparative with the aim of creating an alternative perspective on PMMs. This comparative should help identify and explain new and existing factors that impact the effectiveness of PMMs in achieving project success and suggestions for addressing these factors.

The results of the prestudy—the natural-science comparative—is found in a chapter in the book by Drouin, Müller, and Shankaran (2013) on project management research methods entitled *Novel Approaches to Organizational Project Management Research*.

1.1.2 Project Failure Rates and the Need for Effective PMMs

Projects are the lifeline of an organization's future and are also the truest measures of an organization's intent, direction, and progress (PMI 2013a). Organizations grow and evolve through projectification (Maylor et al. 2006), in which every euro/dollar invested should take the organization one step closer to its stated goals. However, project success rates are low and not improving (Bloch, Blumberg, and Laartz 2012; GOA [Government Accountability Office] 2013; The Standish Group 2010), despite the fact that the knowledge associated with project success and failure has been increasing steadily over the years.

Project failure is estimated yearly in the hundreds of billions of dollars (McManus and Wood-Harper 2008), where failure is not limited to any specific industry (Flyvbjerg, Bruzelius, and Rothengatter 2003; Nichols, Sharma, and Spires 2011; Pinto and Mantel 1990).

To address low project success rates, the project-related knowledge based on research and practitioner experiences has been codified and updated into what are now established standards, PMMs, and guidelines with tools, techniques, processes, and procedures (Flyvbjerg et al. 2003; Morris et al. 2006; Pinto and Mantel 1990).

Lessons learned and ongoing research are continually enhancing PMMs to ensure that success factors are reflected either directly or indirectly within the PMM, guidelines, processes, and procedures (Cooke-Davies 2004). Research has shown that projects that use PMMs provide more predictable project management outcomes than projects that do not use a PMM, but they still suffer from high failure rates (Lehtonen and Martinsuo 2006; Wells 2012).

The literature on PMMs is somewhat contradictory. For example, the literature is split on whether PMMs directly contribute to the goals (Cooke-Davies 2002; Fortune and White 2006; White and Fortune 2002) or to the perceived appropriateness of project management (Lehtonen and Martinsuo 2006). The literature is also divided on whether PMMs that are standardized (Crawford and Pollack 2007), customized, or a combination of both (Milosevic and Patanakul 2005) lead to greater project success. A third view is whether international PMMs (McHugh and Hogan 2011) versus in-house PMMs (Fitzgerald, Russo, and Stolterman 2002) lead to greater project success. Lehtonen and Martinsuo (2006) sum up the research PMMs by stating, "The confusion in research results is reflected also in companies' swing between standardized and tailored systems, and between formal and chaotic methodologies."

The literature covering PMMs, including the divergent views of what constitutes an effective PMM, can be divided into two categories: one that covers PMMs as a homogeneous entity and the other that considers only one part or element of a PMM (e.g., project scheduling or stakeholder management). The term *PMM* implies a homogeneous entity; however, it is really a heterogeneous collection of practices that vary from organization to organization (Harrington et al. 2012). Looking at a PMM as a single entity or an isolated element of a PMM precludes the ability to understand the impact of the interaction of the PMM's elements, which all should contribute to project success. The symptoms of not understanding the building blocks of a PMM and their impact on project success is highlighted in Fortune, White, and Jugdev's (2011) longitudinal study, which found that using PMMs produced a number of undesirable side effects. In this study, 46% of the respondents reported negative side effects. Could these undesirable effects be a consequence of limited research on the interactions among PMM's elements or perhaps missing PMM elements?

To understand the impact of the relationship between a PMM and project success, the building blocks of a PMM need to be understood. As the building blocks of a PMM are not defined, agreed upon, or commonly accepted, the following definition is used for this study: "The building blocks of a PMM are PMM elements that may include processes, tools, techniques, methods, capability profiles, and knowledge areas." A PMM should also take into account the different levels of scope and comprehensiveness, where the term *comprehensiveness* is taken to mean, "Including or dealing with all or nearly all elements or aspects of something" (OxfordDictionaries 2014).

Each organization must decide on the level of PMM comprehensiveness, wherein the more comprehensive the PMM, the less need for it to be supplemented with PMM elements when it is applied to a project. It is unclear from the literature (1) if comprehensive PMMs or PMMs that need to be supplemented lead to greater project success, or (2) what the influence is of project context, notably project governance, on the relationship between project success and a comprehensive PMM or a PMM that needs to be supplemented.

The next section will address project governance and its influence on PMMs.

1.1.3 Governance (Project Governance) as an Environmental Variable

Governance influences organizations, in that it "provides the structure through which the objectives of the organization are set" (OECD 2004). Governance influences people indirectly through the governed supervisor and directly through subtle forces in the organization (and society) in which they live and work (Foucault 1980). Governance in the area of projects takes place at different levels at which there is project governance on individual projects—namely, "the use of systems, structures of authority, and processes to allocate resources and coordinate or control activity in a project" (Pinto 2014, p. 383).

Project governance has been referred to as "the conduct of conduct"; a form of self-regulation in which "the regulator is part of the system under regulation" (Müller 2009, p. 1). Governance influences the way projects are set up (Turner & Keegan, 2001, their organizational structure (Müller, Pemsel, and Shao 2014a), the running of projects (Winch 2001), and their risks strategies (Abednego and Ogunlana 2006). Because governance influences organizations, as well as multiple aspects of project management, it is also likely to influence the value created by project management, especially the effectiveness of a PMM and its impact on project success. The literature does not cover the direct influence of project governance on a PMM or project success, nor does it cover the impact on the nature of the relationship between a PMM and project success. There is a knowledge gap in the literature that is addressed in this research.

Project governance is used in the first and second parts of the main study as the *moderator* (environmental) variable and in the third part of the main study as the *independent* variable.

1.1.4 Subjective Nature of Project Success

Project success is one of the most researched topics in project management because of the importance in understanding what success is and which factors contribute to success (Ika 2009). Despite this, the meaning of the term *project success* is subjective (Judgev and Müller 2005). To achieve a common understanding of project success, it needs to be measurable and, therefore, defined in terms of success criteria (Müller and Turner 2007b). Success criteria are the measures used to judge the success or failure of a project; these are dependent variables that measure success per Morris and Hough (1987). Over the past 40 years, project success factors have been the focus of many researchers (Belassi and Tukel 1996; Cooke-Davies 2002; Pinto and Slevin 1988; Tishler et al. 1996; White and Fortune 2002).

Payne and Turner (1999) define project success factors as, "elements of a project, which, when influenced, increase the likelihood of success; these are the independent variables that make success more likely."

Schultz, Slevin, and Pinto (1987) suggested that the relative importance of success factors varies over the project life cycle, so detailed planning would not be very useful if performed at the end of a project. Success factors are not limited only to a *project* life cycle, they extend into the *product* life cycle as well. Shenhar et al. (2001) described the importance of success factors in both project and product life cycles from project completion to production, and extended that out to preparation for product/service replacement.

Researchers soon realized that success factors without structure, grouping, and context result in increased project risks; therefore, success factor frameworks were introduced (Judgev and Müller 2005). Pinto developed a success framework covering organizational effectiveness and technical validity (Pinto and Slevin 1988). Freeman and Beale's (1992) success framework included efficiency of execution, technical performance, managerial and organizational implications, manufacturability, personal growth, and business performance. Shenhar et al. (2001) described how there is no one-size-fits-all; then, using a four-dimensional framework, he both showed how different types of projects require different success factors and described the strategic nature of projects in which project success should be determined according to short- and long-term project objectives.

Success frameworks also extend to how project success is measured. Pinto and Prescott (1988), Shenhar et al. (2002), Hoegl and Gemünden (2001), and Turner and Müller (2006) developed different measurement models for success that are applicable to different types of projects or different aspects of project success.

Project success is the dependent variable used in Studies 2, 3, and 4 of this research (see Sections 4.6, page 46, and 4.7, page 48).

1.2 Research Focus

1.2.1 Aim and Objectives

The overall aims of this research are as follows:

- Improve the understanding of the impact of a PMM (including its elements) on project success and determine if this relationship is influenced by project context represented by project governance.
- Understand the impact of different project governance contexts directly on a PMM and its elements.
- Understand the impact of different project governance contexts directly on project success.

The specific objectives of this research are as follows:

Academic

1. To understand the relationship between a PMM, including its elements, and project success.
2. If the first research objective is met, then to determine how project context, represented by project governance, influences the relationship between a PMM's elements and project success.
3. To understand the relationship between project governance and a PMM.
4. To understand the relationship between project governance and project success.
5. To create an alternative and new research perspective in the form of a natural-science comparative to see if the findings in objectives 1 and 2 can be explained using such a different research perspective, in addition to finding new phenomena with the new comparative.
6. To provide sufficient evidence that the new natural-science comparative method can be used in future research studies to provide alternative perspectives and new insights that may not be possible with current approaches.
7. To understand the role of project governance in influencing the establishment of a PMM.

Practitioner

1. Provide practitioners with the knowledge of which governance environments are likely to impact the completeness of a PPM and will therefore require supplementing to achieve project success.

2. Provide the project management office (PMO) or other PMM designers with information on whether and when to customize their organization's PMM according to the governance paradigm of the section, department, or organization.
3. Highlight to managements who are considering replacing an institutionalized PMM (including ones with derivatives of their main PMM) the importance of understanding project context and how this is reflected in their incumbent PMM so that an informed decision can be taken on how and whether they should replace the incumbent PMM.
4. Highlight to management how some project governance orientations are more correlated to success than others and also to identify these project success dimensions.

1.2.2 Research Questions

There are five research questions: one relating to the prestudy and the other four to the main part of the research.

Prestudy

The prestudy research question is formulated as follows:

1. *How can a natural science perspective be used in understanding social science phenomena where methodology is the social science phenomena under observation?*
   The unit of analysis is the PMM and project outcome.

In the prestudy, the environmental impact is an integral part of the natural-science comparative, meaning the comparative is contingent on the environment. Therefore contingency theory is being used as the theoretical lens for the prestudy.

Main Study

For the first part of the mixed-method research (qualitative), the core research question is formulated as follows:

2. *What is the nature of the relationship between the PMM, including its elements, and project success, and is this relationship influenced by the project environment, notably project governance?*
   The unit of analysis is the relationship between the PMM and project success.

The above research question was used in the qualitative part of the sequential mixed-methods research.

The second part of the mixed-method research (quantitative) refined the research question as follows:

3. *What is the nature of the relationship between a PMM and project success, and is this relationship influenced by project governance?*
   The unit of analysis is the relationship between the PMM and project success.

For the first and second parts of the main study, contingency theory is being used as the theoretical lens to help understand the influence of environmental factors (project context, notably project governance) on the relationship between PMM and project success.

The third part of the main study—the mixed-method research (quantitative)—looks at the impact of project governance on PMM and project success. The following research questions are asked:

4. *What is the relationship between project governance and a PMM?*
   The unit of analysis is the relationship between project governance and PMM.
5. *What is the relationship between project governance and project success?*
   The unit of analysis is the relationship between project governance and project success.

The third part of the main study uses both agency theory and stewardship theory as the theoretical lens.

1.2.3 Delimitations

For the prestudy, data collection was not limited to any specific geographic location, because nature (genotyping and phenotyping) impacts every part of the globe.

For the main study, the mixed-methods research was not fixed to any set country, although the qualitative part of the study included interviews with 19 people in Switzerland, Germany, UK, and the USA. There was no restriction on the industry sectors. For the quantitative study, the respondents represented industries from North America (38%), Europe (24%), Australasia (22%), and other (15%).

1.3 Structure of the Book

- Chapter 1 is the introductory chapter, in which the empirical and theoretical relevance and the research focus of this dissertation are discussed.

- Chapter 2 refers to the prestudy and further discusses and explains key theoretical themes and concepts investigated in Chapter 5. This chapter provides additional insights to the previously examined theories in the studies and concludes with the knowledge gaps and proposed comparative model.
- Chapter 3 refers to the main study and further discusses and explains key theoretical themes and concepts investigated in the second, third, and fourth research studies presented in Chapters 6, 7, and 8. This chapter brings in additional insights to the previously examined theories in the studies and concludes with the knowledge gaps, hypotheses, and proposed research models.
- Chapter 4 discusses the research methodology, including the philosophy, design, and approaches. This chapter is divided for the prestudy into the qualitative and quantitative parts of the research.
- Chapter 5 is an in-depth comparison of methods for conducting research.
- Chapter 6 examines the effect of methodology on project success.
- Chapter 7 examines the relationship between project methodology and project governance.
- Chapter 8 examines the effect of governance on project success.
- Chapter 9 discusses the use of triangulation to identify interesting phenomena.
- Chapter 10 reviews the research findings and their theoretical implications.
- Chapter 11 discusses the practical implications of the research.
- The appendices contain an overview of the interviews conducted for the qualitative research and the questionnaire used to conduct the quantitative research.

Chapter 2

Use of Comparatives— The Basis for the Natural- Science to Social-Science Comparative

This chapter provides a literature review of key concepts used in the prestudy of the research. It starts with a description of how comparatives are made, describes the concepts within the natural-science comparative, and finally introduces the comparative model.

2.1 Key Concepts

2.1.1 Comparatives

One of the most powerful tools used in intellectual enquiry is comparison, because any observation made repeatedly gives more credence than a single observation (Peterson 2005). Boddewyn (1965) describes comparative approaches as those concerned with the systematic detection, identification, classification, measurement, and interpretation of similarities and differences among phenomena. The disciplines, such as social science (including project management), usually rely on observation rather than experimentation, unlike the natural sciences, for which randomized experiments are the ideal approach for hypothesis testing. However, some research problems cannot be readily addressed using experiments—for example, when looking at research involving two or more species in evolution, ecology, and behavior (Freckleton 2009).

Comparative approaches have been used for years to address the limitations of experiments; virtually every field in biological sciences uses comparatives (Gittleman & Luh 1992). Comparative analysis, unlike experimental studies, has historically relied on simple correlation of traits across species. Over the past 20 years, improvements to comparative frameworks have been made in classifications and the use of statistical methods to the degrees of relatedness in the comparative (Harvey and Pagel 1998; Martins and Garland 1991).

2.2.2 Natural- to Social-Science Comparatives

Comparatives have been made between natural to social sciences using metaphors, such as in the book, *Images of Organization* (Morgan 1997); biological comparatives such as cells of an organism to organizational knowledge (Miles et al. 1997); or behavioral characteristics of a group of organisms called *complex adaptive systems* (CAS) with organizational leadership (Schneider and Somers 2006). Few have gone beyond the juxtaposition, yet still have provided new insights into explaining phenomena that may not have been discovered or explained without these comparatives.

Discussions about the appropriateness of natural or social science approaches to research in projects and their management often refer to the context independence of natural science research. A frequently drawn conclusion is that all social phenomena (such as projects) are context dependent; and therefore, natural science research approaches are deemed inappropriate for gaining understanding of social phenomena (e.g., Flyvbjerg 2001). This perspective may be appropriate in some research studies but presents an oversimplification in others. A great deal of natural science research takes place in context-dependent situations, just as social science research takes place in situations of contextual independence (Knorr-Cetina 1981).

In the field of project management research, comparatives are made mainly through theoretical lenses such as complexity theory, agency theory, stewardship theory, critical point theory, prospect theory, contingency theory, and complex adaptive systems theory. Some of these theories are derived by observing nature, such as complexity theory and complex adaptive systems theory (Brown and Eisenhardt 1997; Holland 2012). Comparatives are performed between two items of interest that may not have been researched—for example, project managers and career models (Bredin and Söderlund 2013).

2.2.3 The Comparative Model

From the literature, there is clearly a need and a benefit in using comparative approaches in the field of project management. A great deal of the man-made world is based on nature and its evolutionary principles, including insights gained by comparing species or comparing a part of an organism, such as a cell or a gene, with the phenotype and behavioral characteristics of that organism.

Dawkins (1974) stated that, "Biology is the study of complicated things that give the appearance of having been designed for a purpose" (p. 1). Project management can be inherently complex in terms of achieving desired outcomes within volatile environments. There are many similarities between biology and project management in terms of complexity, design, impact of changing environments, and product lineage.

From the literature review, we can conclude that there is a knowledge gap when comparing the core makeup and characteristics of an organism with the core makeup and characteristics of project management.

Creating a new comparative, as with any other type of analysis, requires that the phenomena be compared and abstracted from a complex reality. For research, it is important to provide a focus, careful delineation of the scope, the use of defined and accepted terms, and the development of assumptions (Boddewyn 1965).

The focus of the social-science comparative is a PMM. The idea is that the core makeup of a project is based on its used or lived PMM. In the natural science world, the core makeup of an organism is its genotype, which is the genetic makeup of a cell or an organism. One cannot see a genotype, but what can be seen is the organism that is called a *phenotype* (from the Greek *phainein*, "to show" + *typos*, "type"), which is the composite of an organism's observable characteristics or traits. Comparatively, a project applies a PMM, consisting of a set of elements for potential use (genotype); however, in any given project, only the outcome from the use of the applied (sub)set of elements used in a particular project (phenotype) can be seen.

The comparative model is shown in Figure 2.1.

It comprises two levels:

- Level 1 is at the genotype and PMM level, where the latter has been called *progenotype,* for "project genotype."
- Level 2 is at the phenotype of an organism and project outcome. Detailed mapping tables were created in order to build both levels of the comparative. Both the progenotype and project outcome—that is, the project or service—are reified in the comparisons, which helps to make abstract concepts more concrete or real by showing them in a different perspective.

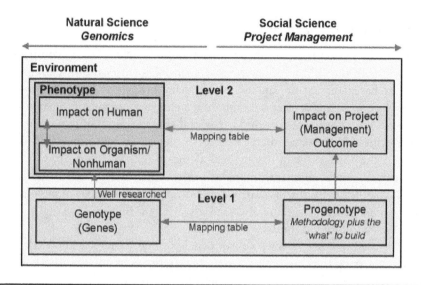

Figure 2.1   Two-Level Comparative Model

The reason for selecting the genotype as a comparative against the progenotype in the social science world is that it allows ideas to be considered that would not be obvious using other methods, and different methods reveal different aspects of a phenomenon.

One of the ideas is described below, which provides an introduction into the thinking behind the comparative. The genes of an organism exist within the DNA of each cell of an organism (Csete and Doyle 2002). Each one of the billions of cells is preprogrammed with information as to what to build and how to build it without any central control. Taking this idea across the social science world of projects is equivalent to having a PMM that contains how and what to build in every element of the PMM. This may make sense, but only in certain types of projects. Referring to Figure 2.2, based on the author's observations, projects are categorized between (1) evolutionary and revolutionary, and (2) one-off projects and ongoing projects. The projects that are more evolutionary than revolutionary would be potentially appropriate to investigate if the idea of combining the "what to build" and "how to build it" into one integrated PMM is beneficial. The other project types would keep the information of what and how to build separately.

Humans have always been inspired by nature, and the majority of inventions throughout our existence have been copied from nature (Vincent 2001).

Would it be possible to learn not only from the outcome (i.e., the phenotype), but how it was created (i.e., the genotype) and the process of organic growth? Figure 2.3 (page 16) shows how man is inspired by birds to create aircraft.

	One-off Projects	Projects to develop Product/Service which will evolve over time	
Evolution	Generic methodology that may or may not tailored to the type* of project	Knowledge of what to build integrated into methodology elements of how to build	Closest to Nature
Revolution	Generic methodology that is tailored to the type* of project	Proven methodology based on a previous product or service that is tailored to the type* of project	

Project types – maintenance, development, research which can result in either a one-off or a product/service ongoing development

Figure 2.2   PMM Approaches for Evolutionary–Revolutionary Project Outcomes

What about creating an integrated PMM for the projects that produce evolutionary products or services, which includes every aspect of what is going to be built and how it is to be built, just like a cell within an organism?

In conclusion, social science (including project management) relies heavily on observation, and comparatives provide a great deal of insight. Humans have been copying nature and, in doing so, have created hundreds of nature-inspired inventions (Benyus 1997) called *biomimicries* (Bar-Cohen 2006). This makes a lot of sense, as nature has perfected each attribute of an organism over thousands of generations.

There are also many similarities between project management and biology, especially epigenetics, which is the study of stable alterations in gene expression-potential that arise during development and cell proliferation (Jaenisch and Bird 2003). Epigenetics has raised a number of questions as to whether humans can not only learn a great deal from the organism (i.e., the phenotype), but can also learn from its genotype and how it evolves. When a project outcome, product, or service can be compared to a phenotype, then can a project's PMM be compared against a genotype, and, if so, what can we learn? The results of the prestudy are described in Chapter 5 and used in the discussion found in Section 10.5 (page 201) to help provide an alternative perspective to the findings in the main part of the research. This is the basis on which the comparative was built.

The next chapter is a literature review of project success, PMMs and the use of contingency theory, and a theoretical lens for the overall study.

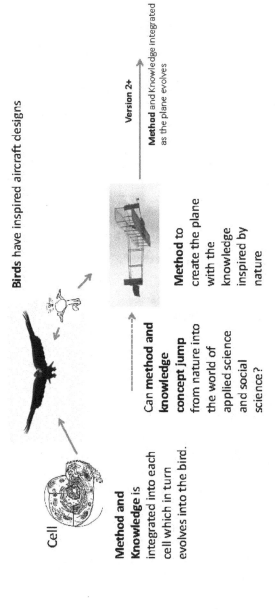

Birds have inspired aircraft designs

Cell

Method and Knowledge is integrated into each cell which in turn evolves into the bird.

Can **method and knowledge concept jump** from nature into the world of applied science and social science?

Method to create the plane with the knowledge inspired by nature

Version 2+

Method and knowledge integrated as the plane evolves

Figure 2.3 Humans Inspired by Nature

Chapter 3

Project Management Methodologies, Project Success, Project Governance, Contingency Theory, Agency Theory, and Stewardship Theory

This chapter provides a literature review of key concepts used in this research. It starts with a discussion of project success, followed by a discussion of PMMs, including differences between a method and PMM; then project governance, first as a context factor and second as an independent variable, is explored; and finally, contingency theory is proposed as the theoretical lens for the first two parts of the main study, with agency and stewardship theory proposed as the theoretical lens for the third and final part of the main study.

3.1 Project Success

Project success is one of the most researched topics in project management because of the importance of understanding how to define success and what factors contribute to achieving it. Despite this, the term *project success* still remains diffuse and, often, in the eye of the beholder (Judgev and Müller 2005).

The measures used to judge the success or failure of a project, called *success criteria,* are the dependent variables that measure success, per Morris and Hough (1987). Defining and agreeing upon project success criteria to make project success measurable is a way to overcome the subjective interpretation of project success (Müller and Turner 2007b).

The understanding of project success has evolved over the past 40 years of research from the simplistic triple-constraint concept, known as the *iron triangle* (time, scope, and cost), to something that encompasses a multidimensional concept comprising many more success criteria attributes (Atkinson 1999; Judgev and Müller 2005; Müller and Judgev 2012; Shenhar and Dvir 2007). Project success is a multidimensional construct that includes both the short-term project management success efficiency and the longer-term achievement of desired results from the project—that is, effectiveness and impact (Judgev, Thomas, and Delisle 2001; Shenhar, Levy, and Dvir 1997).

Even with a concerted effort to define and measure project success, many studies and reports conclude that many projects fail to meet their objectives (Bloch, Blumberg, and Laartz 2012; Cicmil and Hodgson 2006; GOA [Government Accountability Office] 2013; The Standish Group 2010).

Understanding how to measure success is one thing; but if success is rarely achieved, then the focus needs to be on success factors—that is, what needs to be in place for a project to succeed. Project success factors have been the focus of many researchers (Belassi and Tukel 1996; Cooke-Davies and Arzymanow 2003; Pinto and Slevin 1988; Tishler et al. 1996; White and Fortune 2002). To ensure a common understanding of the term *success factors,* the definition from Turner (2008) is used: "Project success factors are elements of a project, which, when influenced, increase the likelihood of success; these are the independent variables that make success more likely."

Some project management literature refers to success factors as *critical success factors* (Lehtonen and Martinsuo 2006; Müller and Judgev 2012; Pinto and Mantel 1990), while other literature refers to them only as success factors (Cooper 1999; Mir and Pinnington 2014). It is useful to understand the origins of the term *success factors* and whether there are any differences between these terms. The concept of success factors was created by Daniel (1961) and refined by Rockart (1979), when Rockart introduced the word "critical" into the term. Rockart described critical success factors as the few key areas in which "things must go right" for the business to flourish. When results in these areas are not adequate, the organization's efforts for the period will be less than desired. Rockart's focus on critical success factors was at the C-level and top management. The term *critical success factors* was later adopted in project management but is not consistently used in the literature. The project management literature referring to critical success factors and success factors overlaps, suggesting that they mean the same thing; therefore, it is assumed as such in this literature review.

Judgev and Müller (2005) carried out a retrospective review on the understanding of project success and found that the number of success factors that have been identified are increasing and also have a longer-term perspective. This can be seen in Figure 3.1, which shows that the literature on critical success factors originally only covered the project execution phase and a small part of the handover phase. Then, over time, the literature extended out to include both the project life cycle and then the product life cycle.

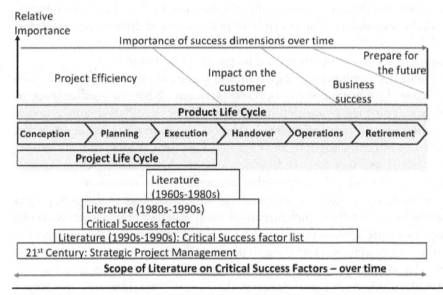

Figure 3.1 Importance of Success Dimensions Over the Project/Product Life Cycles Mapped to Scope of Success Factor Literature (*Source*: Adapted from Judgev and Müller [2005] and Shenhar et al. [2001])

Schultz, Slevin, and Pinto (1987) suggested that the relative importance of success factors varies over the project life cycle. Shenhar et al. (2001) described the importance of success factors, not just on the *project* life cycle but also on the *product* life cycle, from project completion to production, and then to preparation for project/service replacement. The literature on critical project success factors mapped to the importance of success dimensions reveals that it has taken over 30 years to fully understand the implications of the project success dimensions on efficiency, customer impact, business success, and the ability to prepare for the future.

Researchers soon realized that success factors without structure, grouping, and context would result in suboptimal results; therefore, success factor frameworks were introduced (Judgev and Müller 2005). The first integrated frameworks came in the 1990s, at which time Morris and Hough (1987) were pioneers in developing a comprehensive framework on the preconditions of project success (Judgev and Müller 2005). The frameworks varied depending on what aspects were covered. Pinto developed a success framework on organizational effectiveness, technical validity, and organizational validity (Pinto and Slevin 1988). Freeman and Beale's (1992) success framework included efficiency of execution, technical performance, managerial and organizational implications, manufacturability, personal growth, and business performance. Shenhar et al. (2001) described how there is no one-size-fits-all; then, using a four-dimensional framework, he showed how different types of projects require different success factors and described the strategic nature of projects on which project success should be determined according to short- and long-term project objectives.

Figure 3.1 combines the importance of success dimensions over time with the scope of the literature on critical success factors. What is immediately apparent is the short-term view that researchers took in the 1960s–1980s in understanding how projects are executed; then the literature expanded in terms of a greater short-term understanding of project efficiency by looking at the project life cycle as well as a forward-looking view, which also encompassed the later phases of the product life cycle—that is, operations and retirement.

In summary, project success can mean different things to different project stakeholders; therefore, implementation of measurable project success criteria helps to ensure agreement and understanding. Determining whether a project has met its objectives is one thing, but when attention is not paid to determining what the relevant project success factors are, project success is unlikely to be achieved. As research continues on the topic of project success, new project success factors are being identified that take a longer-term perspective and are being added into the success factor frameworks.

Project success is the dependent variable in the research models of the main study.

3.2 Project Management Methodologies (PMMs)

The use of the terms *project methods* and *PMMs* are sometimes confused (Avison and Fitzgerald 2003); therefore, definitions are provided below to ensure a common understanding for the reader.

The term *methodology* is derived from Greek *methodologia* and is defined as: "A system of methods used in a particular area of study or activity" (Oxford Dictionaries 2014).

In the project management field, *project management methodology* (PMM) is defined as: "A system of practices, techniques, procedures, and rules used by those who work in a discipline" (PMI 2013a).

A collection of procedures, techniques, tools, and documentation aids will help system developers in their efforts to implement a new information system. A PMM consists of phases and subphases, which will guide system developers in their choice of techniques that could be appropriate at each stage of the project and also help them plan, manage, control, and evaluate information systems projects (Fitzgerald, Russo, and Stolterman 2002).

The term *method* is derived from Greek *methodus* and is defined as: "A particular procedure for accomplishing or approaching something, especially in a systematic or established manner" (OxfordDictionaries 2014).

Forty years ago, the first formal PMMs were set up by governance agencies to control project budget, plans, and project quality (Packendorff 1995). Since that time, the literature on PMMs has covered standardization, customization, or a combination thereof; in-house, international methodologies; soft factors; the roles of the PMO in PMM development; and the impacts of PMMs on project success.

3.2.1 Customization

Turner and Cochrane (1993) and Shenhar and Dvir (1996) were a few of the first proponents of customization to show that projects exhibit considerable variation, which at the time went against the literature trend that assumed all projects were fundamentally similar. Payne and Turner (1999) found that project managers often report better results when they can tailor procedures to the type of project they are working on, matching the procedures to the size of the project or the type of resource working on the project. Wysocki (2011) stated that the often-used term *one size fits all* does not work in project management.

McHugh and Hogan (2011) found that in-house PMMs work well, but that there are demands from external customers for a recognized PMM. This is, in part, due to the assurance that the organization is using what is considered to be "best practices" as well as a supply of trained resources.

3.2.2 Standardization

PMMs and processes have been referred to as *organizational processes*, implying that they have degrees of standardization (Curlee 2008). However, there is a risk that structured methodologies, when developed in a normative way, become prescriptive and are based on a series of checklists, guidelines, and mandatory reports (Clarke 1999). The ISO 9000 standards are frequently criticized for being too standardized and prescriptive while generating excessive costs and paperwork (Brown, Wiele, and Loughton 1998; Stevenson and Barnes 2001). Crawford (2006) found that "owners" of project management practices were following a path of corporate control and standardization, whereas the project managers showed that certain project processes did not apply to their local projects, thus creating a tension between project managers and the corporate control and related standards. The primary function that promotes standardized methodologies is the project management office (PMO). Hobbs, Aubry, and Thuillier (2008) observed the dilemma that exists between PMOs that are focused on standardizing organizational PMM and the need for flexibility in the execution of a project.

3.2.3 Combination of Standardization and Customization

A contingency approach to standardizing parts of a PMM was suggested by Milosevic and Patanakul (2005), in which it made sense to standardize only parts of the PMM in an organization.

In summary, there is little consensus in the literature on whether PMMs should be standardized, customized, or a combination of both. Aubrey et al. (2010) found that the more experienced PMOs are using new methods derived from agile methodologies that allow flexibility in processes and PMMs. This suggests a contingency between PMMs and project success; however, the literature is also split on whether PMMs directly contribute to the goals (Aubry et al. 2010) or to the perceived appropriateness of project management (Lehtonen and Martinsuo 2006).

The description of a PMM varies among the international PMM standards. For example, the Project Management Institute (2013a) describes a PMM as a system of practices, techniques, procedures, and rules; whereas Prince II is not described as a PMM but rather as a method (OGC 2002) that contains processes but requires techniques to be added. Ericsson (2013) considers its PROPS PMM to be a model and not a PMM, wherein the model describes all of the project management activities and documentation. Anderson and Merna (2003) have helped to categorize PMMs into process models, knowledge models, practice models, and baseline models. The categorization of PMMs helps in understanding what type of project/industry the PMM is targeting, but it does not help in understanding the impact of a PMM on project success.

Perhaps it is not sufficient to look at the PMM as a whole, but instead at its building blocks, which the author terms PMM *elements*. If the elements of a PMM are understood, then this is the foundation for understanding how the elements collectively impact project success. PMMs comprise a number of heterogeneous elements that, when applied to a project, should have a positive impact on project success.

To achieve the desired effect, a PMM needs to take into account scope and comprehensiveness, where the term *comprehensiveness* is defined as including or dealing with all or nearly all elements or aspects of something (OxfordDictionaries 2014). PMMs that are not comprehensive are considered incomplete in this study and, therefore, will need to be supplemented during the project life cycle.

From a survey of project management current practices, White and Fortune (2002) found that very few methods, tools, and techniques were used; and for the ones that were used, almost 50% of the respondents reported drawbacks in the way they were deployed. This suggests that organizational PMMs suffer from a lack of applicability and comprehensiveness. Research supports this assessment, as organizations experience limitations in their PMMs irrespective of whether they are in-house or off the shelf (Fortune et al. 2011; Joslin and Müller 2016; Wells 2013). When the selection of PMMs was carried out at the organizational level, it frequently did not address the needs of the departments and projects, resulting in project managers tailoring their organizational PMMs specifically for their projects (Wells 2013).

In summary, there is no agreement as to what makes up a PMM, which may influence whether or not a PMM is seen as being comprehensive. Researchers are also divided as to whether standardized versus customized methodologies contribute more or less to project success, but observations have been made in which experienced PMOs are applying a contingency approach to PMMs. There is also the question of whether PMMs contribute directly to project success or indirectly via the appropriateness to project management.

The reason for selecting a PMM as the independent variable in the research model is to determine whether a PMM directly or indirectly contributes to project success; if directly, what percentage of project success can be accounted for by correctly applying a PMM; and is the relationship between a PMM and project success influenced by environmental factors, notably project governance.

3.3 Project Governance

The reason for considering project governance as the context factor is that corporate governance, including the governance of projects and project governance, is present from the point of the creation of an organization. Governance has influenced the way individuals have viewed project management because it provides the structure through which projects are set up, run, and reported (Turner 2006). Therefore, project governance is also likely to influence the choices taken in selecting, applying, and evolving a PMM and the relationship between a PMM and project success.

The Office of Government Commerce (OGC) (2002) terms governance as a framework that defines the accountability and responsibility of people who are driving the organization as well as the structure, policies, and procedures under which the organization is directed and controlled. Governance theory was originally developed from policy research in political science (Friedmann 1981; Krieger 1971; Nachmias and Greer 1982) and extended to encompass many levels, including international governance, national governance, corporate governance, and project governance (Klakegg, Williams, and Magnussen 2009). Researchers have addressed governance from different perspectives, using a number of theories to help explain observed phenomena. The most notable theories are shown in Table 3.1.

Table 3.1 List and Description of Governance Theories

Theory	Description	Key authors
Agency theory	Agency relationship exists between two parties (the principal and the agent) in organizations where both actors are perceived as rational economic actors who act in a self-interested manner.	Mitnick (1973), Jensen and Meckling (1976)
Stewardship theory	Actors (managers) are stewards whose motives are aligned with the objectives of their principals rather than their own goals.	Donaldson and Davis (1991)
Transaction cost economics	Organizations achieve the lowest transaction costs to produce a product or service and adapt their governance structures to achieve this.	Williamson (1979)
Stakeholder theory	Actors (managers) focus more on the stakeholders than the shareholders.	Donaldson and Preston (1995)
Resource dependency theory	Directors and managers are able to prioritize, acquire, and facilitate the utilization of resources aligned to organizational objectives.	Pfeffer and Salancik (1978)
Contingency theory	Change in the effect of one variable (an independent variable) on another variable (a dependent variable) depending on some third variable (i.e., the moderator variable).	Donaldson (2001, p. 5)

The use of contingency theory as the theoretical lens in studies 1, 2, and 3 and agency theory and stewardship theory as the theoretical lens in study 4 are described in detail in Sections 3.4 and 3.5.

Governance in the area of projects takes place at different levels at which there is project governance on individual projects—namely, "the use of systems, structures of authority, and processes to allocate resources and coordinate or control activity in a project" (Pinto 2014, p. 383). There is governance for groups of projects, such as programs or portfolios of projects, in which the emphasis is on collective governance, which is viewed as governance of projects (Müller and Lecoeuvre 2014).

Project governance is defined by the Project Management Institute as, "The alignment of project objectives with the strategy of the larger organization by the project sponsor and project team [. . .] is defined by and is required to fit within the larger context of the program or organization sponsoring it, but it is separate from organizational governance" (PMI 2013b, p. 553).

The governance of projects combined with project governance coexist within the corporate governance framework and cover project portfolio, program, and project management governance (Müller, et al. 2014).

The literature on project governance models or guidelines addresses different contexts, such as project governance for risk allocation (Abednego and Ogunlana 2006), a framework for analyzing the development and delivery of large capital projects (Miller and Hobbs 2005), NASA-specific framework for projects (Shenhar et al. 2005), governing the project process (Winch 2001), mechanisms of governance in project organizations (Turner and Keegan 2001), normalization of deviance (Pinto 2014), stakeholder management (Aaltonen and Sivonen 2009), project governance roles (Turner 2008), and governance in project-based organizations (functional, matrix, or projectized) (Müller et al. 2014).

One can conclude from the literature that project governance is contingent on its application and also on its environment. The literature does not cover understanding the impact of project governance on the relationship of PMMs and project success, which is addressed in this research.

To understand the impact of governance on the relationship between PMM and project success, a method to categorize each organization's governance positioning is required, ideally within a governance model.

Governance models are developed from different perspectives using either a top-down or a bottom-up approach (Klakegg et al. 2009). Top-down approaches are developed from a shareholder-outcome perspective, whereas bottom-up approaches take a process-control perspective and can be considered as an extension of a PMM (Müller 2009). The present study requires a governance model that considers the perspectives of shareholder versus stakeholder and a "follow the process" versus "get it done" (outcome) approach. This is because the governance model perspectives map to the overall objective of a project—that is, a successful outcome—with the objective of a PMM (structured approach to deliver a project), all within an environment that is influenced by shareholders and stakeholders.

The governance models that incorporate topics such as ethics, corporate citizenship, roles, and responsibilities (Dinsmore and Rocha 2012; Renz 2008; Turner 2008; Walker, Segon, and Rowlingson 2008) were excluded because the emphasis of the study is on the shareholder–stakeholder and behavior–outcome aspects of the organization. Therefore, the most relevant model was Müller's governance model (2009), which draws on the theories of transaction cost economics, agency theory, and institutional theory, using legitimacy to emphasize conformance.

Müller's governance model explains the governance of projects through four governance paradigms, where one paradigm is used for one project by the project's parent organization; however, the use of paradigms can vary across the organizational units throughout the organization.

The governance paradigms from Müller (2009) are shown in Figure 3.2, along with the theories that the paradigms are derived from.

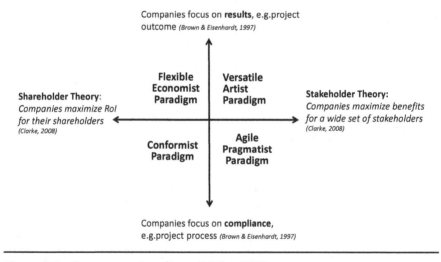

Figure 3.2   Governance Paradigms (Müller 2009)

Müller's governance model addresses corporate governance orientation and control orientation at the level of the organizational unit that governs a project. The corporate governance dimension builds on the Clarke (1998) and Hernandez (2012) models, which claim that a corporation's governance orientation can be found on a continuum from a shareholder to a stakeholder orientation:

- *Shareholder theory* of corporate governance assumes that the main purpose of an organization is to maximize shareholder return (Brown and

Eisenhardt 1997). Therefore, the value system of these types of organizations prioritizes shareholders over stakeholders, and qualitative objectives, such as employee well-being, good relationships with interest groups, and ethical standards, take second priority (Müller 2009).

- *Stakeholder theory* takes the wider social responsibility of organizations into account. An organization that is stakeholder oriented balances the qualitative and quantitative requirements of a wide range of stakeholders (Müller 2009). The purpose of the organization is to create wealth and value for its stakeholders (Clarke 1998).

Referring again to Figure 3.2, the corporate governance orientation is the horizontal line along which pure shareholder or stakeholder orientation exists at opposite ends of the continuum. Every organization will be found somewhere on this continuum.

The second dimension is about control, representing the control exercised by the governing institution over the project and its manager. There is a distinction between organizational control, which focuses on goal accomplishment by controlling outcomes (e.g., reaching a set of objectives) versus compliance with a focus on employees' behaviors (e.g., following a process, such as a project management PMM) (Brown and Eisenhardt 1997; Ouchi 1980; Ouchi and Price 1978).

The vertical line in Figure 3.2 is the control orientation in which the extreme points are organizations focused solely on goal accomplishment by controlling the results (outcomes) versus compliance in employee behavior (following a process or a project management PMM). Every organization will be found somewhere on this continuum.

Each of the quadrants in Figure 3.2 represents a governance paradigm, and every organization can be represented by one of these paradigms. The names and attributes of each paradigm are shown in Figure 3.3.

Using the governance paradigms from Müller (2009), including the scales, governance is the moderating variable in the research models for Studies 2 and 3 and the independent variable for Study 4.

3.4 Contingency Theory—Theoretical Lens for the Prestudy and First Two Parts of the Main Study

A management theory developed more than 50 years ago, called *contingency theory*, suggested that there is no single best way to manage and structure an organization (Burns and Stalker 1961; Woodward, Dawson, and Wedderburn 1965). Woodward (1965) suggested that technologies influence organizational attributes such as span of control, centralization of authority, and the development of rules and procedures. Burns and Stalker (1961) introduced the concepts of mechanistic

	Shareholder Orientation	*Stakeholder Orientation*
Outcome control focus	**Flexible Economist Paradigm** • Highest possible Return on Investment (ROI) • Project management as core competence • Professional project managers • Guided by tactical Project Management Offices (PMO)	**Versatile Artist Paradigm** • Balancing requirements of a wide range of stakeholders • Tailoring of methods • Project management a core competence • Project management a profession • Guided by a strategic PMO
Behavior control focus	**Conformist Paradigm** • Maximizing shareholder return • Project management a subset of development processes for technical products or services. • Project management is understood as on-the-side task	**Agile Pragmatist Paradigm** • Balances the diverse requirements of a variety of stakeholders by maximizing their collective benefits • Maximize value by strict prioritization of user needs.

Figure 3.3   Governance Paradigms, Names and Attributes (*Source:* Müller [2009], © PM Concepts AB, 2009, used with permission)

versus organic organizations and how organic organizations are better suited to dynamic, changing environments. Lawrence and Lorsch (1967) showed that varying rates of change affect the ability of organizations to cope. Since the 1960s, there has developed a large body of literature on structural contingency theory (Argote 1982; Donaldson 1987; Miles and Snow 1978; Ouchi 1980).

So what is the contingency theory of organizations? It is the change in the effect of one variable (an independent variable) on another variable (a dependent variable) depending upon a third variable (the moderator variable) (Donaldson 2001, p. 5).

In the field of project management research, prior to the 1980s, contingency factors were rarely taken into account when researching project management topics (Judgev and Müller 2005). In the late 1980s, the first studies on project management started to use contingency theory as a theoretical lens for project context (Donaldson 2006). Research topics that have used contingency theory include the topology of projects with minor and major impacts (Blake 1978), innovation types in business (Steele 1975), product development project types (Wheelwright and Clark 1992), leadership styles for project and functional managers in organizational change (Turner, Müller, and Dulewicz 2009), project procedures tailored to context (Payne and Turner 1999), leadership styles according to project type (Müller and Turner 2007a), and project type and the ability to select appropriate management methods linked to project success (Boehm and Turner 2004; Shenhar and Dvir 1996).

A recent bibliographical review of the use of contingency theory in the field of project management showed that contingency is increasingly being applied in research papers, with a noticeable increase since 2005 (Hanisch and Wald 2012). Fitzgerald, Russo, and Stolterman (2002) noted that the most successful PMMs are those developed for the industries or organizations that are aligned to the context factors. Lehtonen and Martinsuo's management study of project failure and the role of project management PMMs concluded, "Some contingency variables may have an impact on the relation between PMM and success" (Lehtonen and Martinsuo 2006, p. 10). This supports the notion of contingency theory, in which the independent variable, PMM, and the dependent variable, success, are influenced by a third variable.

Contingency theory is being used as the theoretical lens for Studies 1, 2, and 3 to help understand the impact of PMMs on project success and to determine that governance is acting as a contingency (moderator) variable. The aim of the main part of the research is to understand the impact of PMMs on project success in the context of governance.

The use of contingency theory in the prestudy is implicit, because, in the natural science world of biology, every organism is influenced by its environment (Dawkins 1974). The comparative takes this contingency perspective across to the social sciences in terms of showing how PMMs are also contingent on their project and organizational environments.

3.5 Agency Theory and Stewardship Theory— Theoretical Lens for the Third Part of the Main Study

Agency theory and stewardship theory are two opposing but appropriate theories. Agency theorists argue that corporate managers (agents) may use their control over the allocation of corporate resources opportunistically in order to pursue objectives contrary to the interests of the shareholders (principals) (Jensen and Meckling 1976). Agency theory has been used by researchers in traditional finance, economics, marketing, political science, organizational behavior, sociology, and corporate governance (Eisenhardt 1985). The formal definition of agency theory states that an agency relationship exists between two parties (the principal and the agent) in organizations in which both actors are perceived as rational economic actors who act in a self-interested manner (Mitnick 1973).

Agency relationships are referred to as occurring between two parties— that is, the principle and the agent—but there can be several principle agents in a project process, such as procurement of resources or change request processes (Toivonen and Toivonen 2014). Corporate and project governance, when designed correctly within the context of the organization, should minimize the risks and issues associated with agency issues. Agency theory based on Jensen and Meckling's (1976) view of principle agent models has been criticized because they neglect to consider that the principle–agent transitions are socially embedded and therefore impacted by broader institutional contexts (Davis, Schoorman, and Donaldson 1997a; Wiseman, Cuevas-Rodríguez, and Gomez-Mejia 2012).

Stewardship theory arose in response to criticism regarding the generalizability of agency theory, which states that the actors (managers) are stewards whose motives are aligned with the objectives of their principles, rather than being motivated by their own goals (Donaldson and Davis 1991). The steward differs from the agent in that the steward is trustworthy and will make decisions in the best interests of the organization. This is achieved by meeting the organization's demands as well as the steward's personal needs. The steward aligns the personal and organizational interests by prioritizing the long-term goals over short-term gains (Davis, Schoorman, and Donaldson 1997b). Stewardship theory has been criticized because it views the organization in a static way and does not account for stewards' resorting back to an agent position when their positions are threatened (Pastoriza and Ariño 2008). Neither agency theory nor stewardship theory is more valid than the other, in that each may be valid for different types of phenomena (Davis et al. 1997b).

The third part of the main study investigates the relationship between project governance and project success and project governance and a PMM, wherein observed phenomena will be described through the agency theory–stewardship theory lens.

Chapter 4

Research Methodology

This chapter presents the research philosophy and explains the details of the research model for each of the four studies, describes an integrated research model across all studies, offers a section on philosophical triangulation, and concludes with the research methodology for each of the studies.

4.1 Research Philosophy

The approaches, strategy, choices, time horizons, techniques, and procedures applied in this study are based on a philosophical perspective. This perspective drives the research design and provides the basis for how the results are interpreted (Easterby-Smith, Thorpe, and Jackson 2008). The determinants of good social science are not the methods selected but the underlying ontology and epistemology perspective (Alvesson and Sköldberg 2009).

The research paradigm guides how decisions are made during the research process. Paradigms can be characterized through their ontology (what is reality), epistemology (how do you know something is true), and methodology (approach/process). Combined, these characteristics create a holistic view of how knowledge is understood, how the researcher is positioned in relation to this knowledge, and what methodological strategies were used to discover it (Guba and Lincoln 1994). It is ontology and epistemology, rather than methods, that are the determinants of good social science (Alvesson and Sköldberg 2009).

There are four commonly used versions of research paradigms in social science (Morgan 2007) whose origins are mostly derived from works such as Thomas Kuhn's seminal book, *The Structure of Scientific Revolutions* (Kuhn 1970), and Burrell and Morgan's book, *Sociological Paradigms and Organisational Analysis* (Burrell and Morgan 1979). The three versions of the paradigm concept share the same belief system that influences the kind of knowledge researchers seek and how they interpret the data they collect.

4.1.1 Paradigms as World Views

This first perspective of a paradigm is from the ontological perspective (the science or study). The ontological choices range from positivism to constructivism (Alvesson and Sköldberg 2009; Tashakkori and Teddlie 2009). The research study will use positivism in the prestudy (see Chapter 5), which, from an ontological perspective, assumes objectivity in an external reality, which means the researcher is detached from the subject of research and, therefore, does not influence the phenomena. It is appropriate for the prestudy, because the author maps the attributes of the subjects under observation and, therefore, will not influence the phenomena.

Critical realism serves as the underlying paradigm in the qualitative part of main mixed-methods research. Critical realism emphasizes the existence of an objective reality independent of the researcher's descriptions and ideas (Alvesson and Sköldberg 2009). Within critical realism, social constructions are recognized, but they are outlined in an objectivist way (Alvesson and Sköldberg 2009). Critical realism takes a middle-ground position between positivism and interpretivism. In positivism, theory aims to predict, whereas in interpretivism, theory describes conditions or context for the production of meaningful experiences (Wikgren 2005). Critical realism emphasizes the need to critically evaluate objects to understand social phenomena (Sayer 1992). Critical realism consists of different levels, which addresses the fact that complex social phenomena cannot be explained solely by looking at mechanisms and processes that operate purely on one level (Wikgren 2005). Entities might be analyzed at different aggregation levels, in which some entities also emerge from lower levels (Easton 2010). The qualitative part of the interviews will be carried out at the upper levels, whereas an online questionnaire will be carried out at the lower levels.

Post-positivism is the final paradigm used for the two quantitative studies as part of the mixed-methods research. Post-positivism assumes that an objective and extrinsic reality (facts and laws) exists (Tekin and Kotaman 2013). However, the perspective of post-positivist research is not to establish generalizations about the phenomenon under observation, but rather to focus on the "meaning and understanding of the situation or phenomenon under examination"

(Crossan 2003, p. 54). Project governance and project success are both socially constructed phenomena; therefore, the impact of project governance on project success is investigated to provide conditional knowledge that can be used to understand when and how to improve project governance's positive impact on project success.

4.1.2 Paradigms as Epistemological Stances

This second perspective of paradigms comprises the concepts of knowledge, model, and testability (Bunge 1996). The prestudy will look for facts and causes (positivist), while the main part of the study (see Chapters 6, 7, and 8) will look at relationships between PMM and project success and governance. The use of PMM in the prestudy and the main study will create the possibility for theory and methodological triangulation. This should provide a more holistic understanding than would have been reached using a singular methodology (Tashakkori and Teddlie 2009).

4.1.3 Paradigms as Shared Beliefs in a Research Field

This third perspective of paradigms is from a research methodological viewpoint and is used to design research or a conceptual framework and processes to guide the inquiry of knowledge on the topic of interest. The main part of the research proposes using mixed methods, because the qualitative part will provide a deeper understanding of a methodology and the impacts of the environment on the relationship of a PMM and project success. The finding of the qualitative study (see Chapter 6) will be used to create the online survey for the quantitative part of the mixed methods, where the findings can then be generalized.

4.2 Approach, Strategy, and Choices

4.2.1 Approaches

In research, there are two broad methods of reasoning referred to as the *deductive* and *inductive* approaches (Tashakkori and Teddlie 2009).

Deductive reasoning starts from a general rule and moves to the more specific rule (sometimes called a *top-down* approach) and asserts that, ". . . the rule explains a single case" (Alvesson and Sköldberg 2009).

Inductive reasoning starts from specific observations and moves to broader generalizations and theories (sometimes referred to as the *bottom-up* approach). A number of cases are considered and assume a connection exists, which is generally valid. The inductive approach is used for understanding a new or unknown phenomenon and collects data through interviews, observations, and focus groups (Miles and Huberman 1994).

The prestudy uses a deductive approach, because it is based on the principles of science and uses already established knowledge, which is tested in a new circumstance. The main study uses a deductive approach, which uses existing theory for hypothesis development and testing.

4.2.2 Strategies

The prestudy is based on transformative research by suggesting a particular empirical natural science perspective for some social science phenomena, such as research in project management methodologies and project portfolios as a comparative to existing perspectives. The National Science Foundation (NFS) (2007, p. 10) describes transformative research as:

> . . . research driven by ideas that have the potential to radically change our understanding of an important existing scientific or engineering concept or leading to the creation of a new paradigm or field of science or engineering. Such research also is characterized by its challenge to current understanding or its pathway to new frontiers.

Transformative research results often do not fit within established models or theories and may initially be unexpected or difficult to interpret; further, their transformative nature and utility may not be recognized until years later. Transformative research has the following characteristics: challenges conventional wisdom; leads to unexpected insights that enable new techniques or methodologies; and/or redefines the boundaries of science, engineering, or education.

Why transformative research? Because it is trying to answer the following question:

How can a natural science perspective be used in understanding social science phenomena?

For the main study, the qualitative study is based on interviews to determine the source of the questions and contextual information that helped to formulate the questionnaire in the quantitative part of the study.

4.2.3 Choices

For the prestudy, a conceptual approach was taken using a natural science perspective in an attempt to answer those parts of the research questions that remain unanswered by the main study.

For the main study, a mixed-research method (qualitative [Chapter 6] and quantitative [Chapter 7 and Chapter 8]) was chosen in order to answer the research questions from both the critical realism and post-positivist perspectives. For the qualitative studies, the overall methodological approach of the study was deductive. However, the research model was qualitatively validated through interviews that were inductively analyzed.

4.3 Research Process Sequence

The research process depended in part on the findings from the prestudy. In the event that a relationship between natural science and social science (project management) was not made, this would have impacted the main part of the study. As a consequence, the research question for the prestudy, "How can a natural science perspective be used in understanding social science phenomena?" has a different focus than the main study. Once a link was found, then the following research question was posed in the main study:

What is the nature of the relationship between the PMM, including its elements, and project success, and is this relationship influenced by the project environment, notably project governance?

Figure 4.1 illustrates the steps taken in the research process. The symbol (E) denotes empirical information based on interviews, meetings, and surveys, whereas the symbol (L) denotes documents and theoretical data gained through the literature reviews.

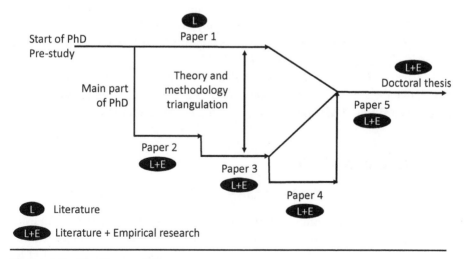

Figure 4.1   The Research Process

With the prestudy being primarily related to natural science, the positivist, deductive approach was taken using transformative research strategy. The prestudy (Chapter 5) that resulted from the research that resulted from the research was based on literature findings and a number of key concepts such as Universal Darwinism (Nelson 2006), gene-centric view of evolution (Dawkins 1974), and evolutional stable society (Dennett 1996). The pre-study contained the findings from the analysis and the mapping tables that were created to compare the attributes of both worlds (natural and social sciences) and a detailed literature review on comparative analysis approaches, as well as a theory-building section that showed that comparatives already exist across sciences (which also comply with Universal Darwinism) and are used in research.

Once the prestudy was completed and the link created between natural and social sciences (methodology), the main part of the PhD could commence. Paper 2 (Chapter 6) was produced based on the literature (L) and findings (E) from the qualitative study. Paper 3 (Chapter 7) used the questions and findings that came out of the interviews in the qualitative research to create an online survey, and the findings from the survey and literature were the basis for Paper 3. Paper 4 (Chapter 8) used the same data collected from the online survey that related to governance and project success to understand if there was a direct relationship between the two variables. Paper 5 (Chapter 9) uses the three philosophical stances of Papers 1 to 4 and proposes using philosophical triangulation to identify interesting new phenomena

This book brings both the prestudy and the main study together in the discussion section.

Paper 1 was presented in June 2014 at the EURAM research conference held in Valencia, Spain, and a revised version was published in the *Project Management Journal*® (*PMJ*). Paper 2 was presented in July 2014 at the PMI research conference in Portland, Oregon, and a revised version was published in the *International Journal of Managing Projects in Business* (*IJMP*). Papers 3 and 4 have been published in the *International Journal of Project Management* (*IJPM*). Paper 5 was presented at the EURAM Research Conference in June 2015 and won the Best Paper Award from PMI and IPMA. It was then published in *IJPM*.

4.4 Research Models

4.4.1 Prestudy—Derived Model—Research Model 1

The concept of genotyping and phenotyping was the basis of the exploratory research. The environment impacts both the natural and social science worlds, so the challenge was to find which aspect(s) of project management could be modeled to the natural science world. This was achieved from reviewing the literature in project management until the idea came that the core makeup of

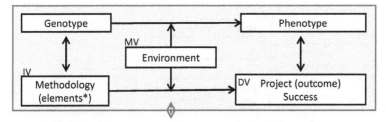

* Elements include processes, tools, techniques, knowledge areas,
  capability profiles, methods

IV – Independent variable
DV – Dependent variable
MV – Moderator variable

Figure 4.2 Prestudy-Derived Research Model

a project is its applied PMM and its elements. Therefore, an applied PMM and its elements can be compared to a genotype comprising genes and the project outcome to a phenotype, where both are impacted by the environment as shown in Figure 4.2.

4.4.2 Qualitative Research Model—Research Model 2

The main part of the PhD used the output of the prestudy to create a focus on PMMs.

The literature on PMMs in Section 3.2 (page 20) describes the importance of PMMs as their integral role within a project but raises questions as to whether PMMs contribute directly to project success or indirectly by means of the appropriateness to project management (Lehtonen and Martinsuo 2006). Research on PMMs focuses either on a part (or element) of a PMM, such as project scheduling or stakeholder management, or as a whole or homogeneous entity—for example, a PMM and the impact on success (Mir and Pinnington 2014).

There are PMMs that are standardized, customized, and a combination of both, implemented in different organizational environments, which indicates there are one or more project environmental factors present (Fortune et al. 2011; Lehtonen and Martinsuo 2006; Milosevic and Patanakul 2005; White and Fortune 2002).

This leads to the following research question:

What is the nature of the relationship between the PMM, including its elements, and project success, and is this relationship influenced by the project environment, notably project governance?

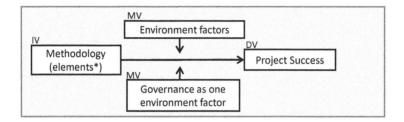

* Elements include processes, tools, techniques, knowledge areas, capability profiles, methods

IV – Independent variable
DV – Dependent variable
MV – Moderator variable

Figure 4.3   Qualitative Research Model

The research model (see Figure 4.3) was derived from the research question, and the subsequently developed propositions were derived from the literature review in Chapter 3:

- **Proposition 1.** There is a positive relationship between project management methodology (PMM) and project success.
- **Proposition 2.** There is a moderating effect of the project environment, notably governance, on the relationship between a PMM and project success.

Note: The research model is shown even though in qualitative research there are normally no research models, because the aim is to understand meanings that people attach to phenomena and not to test variables and their relationships. However, the model is shown because it provides a link to the prestudy and the quantitative part of the study, which is explained in Section 4.4.5 as an integrated research model.

4.4.3 Quantitative Research Model 3

The research model was refined from the qualitative research to focus on project governance as a moderating variable (see Figure 4.4). Governance influences everyone indirectly through the governed supervisor and directly through subtle forces in the organization (and society) in which they live and work (Foucault 1980). Joslin and Müller (2016) found, from a qualitative study, that governance was the most mentioned environmental factor influencing the effectiveness of a

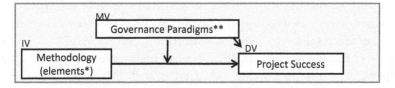

*Elements include processes, tools, techniques, knowledge areas,
capability profiles, methods
** Governance paradigms Müller (2009)

IV – Independent variable
DV – Dependent variable
MV – Moderator variable

Figure 4.4   Quantitative Research Model

PMM. This discovery, combined with the assumption that some form of governance is present in every organization, may have an influence on project selection, set up (including PMM), execution guided by a PMM, and project close, which ultimately could impact project success.

This leads to the following research question:

What is the nature of the relationship between a PMM and project success, and is this relationship influenced by project governance?

Organizations' PMMs vary in completeness and appropriateness from organization to organization, in that some are considered inadequate for certain types of projects (Fortune et al. 2011; Joslin and Müller 2015b; Wells 2012; White and Fortune 2002). The term *comprehensive set of PMM elements* is used in this study to indicate a PMM's completeness and appropriateness for an organizational environment. A comprehensive set of PMM elements includes tools, techniques, processes, methods, knowledge areas, and capability profiles, which will address the needs of the projects within an organization. The difference between a PMM and a comprehensive PMM is that a comprehensive PMM does not need to be supplemented by PMM elements, whereas a PMM may need to be supplemented—that is, it may be insufficient for a given project. The phrase "supplement missing PMM elements" is used to indicate that an organization's PMM has been supplemented by the project manager because the PMM is incomplete or inadequate. The phrase "apply relevant PMM elements" indicates that the project manager, irrespective of whether he or she has supplemented any missing PMM elements, has applied the relevant PMM elements to achieve the expected outcome by applying these PMM elements.

Referring to the literature review on PMMs in Section 3.2, research has shown that projects in which methodologies are used provide more predictable and higher success rates (Lehtonen and Martinsuo 2006; Wells 2012). However, for projects that do use PMMs, there are still high project failure rates (Wells 2012). Wells goes on to say that the selection of PMMs at the organizational level did not address the needs of the departments and projects, and, in some cases, project managers would tailor their organizational PMMs. This implies adding, changing, and removing PMM elements.

Therefore, the following hypothesis is derived.

- **Hypothesis 1.** There is a relationship between the PMM's elements and project success.

Now the following subhypotheses can be defined:

- o **H1.1.** There is a positive relationship between a comprehensive set of PMM elements and project success.
- o **H1.2.** There is a positive relationship between supplementing missing PMM elements and project success.
- o **H1.3.** There is a positive relationship between applying relevant PMM elements and project success.

Referring to the literature in Section 3.2 on PMM, the effectiveness of a PMM is contingent on the environment (Fitzgerald, Russo, and Stolterman 2002; Fortune et al. 2011; Lehtonen and Martinsuo 2006; Shenhar et al. 2002; White and Fortune 2002). Referring to Section 3.3 (page 23) of the literature review, governance influences projects in the way they are set up (Turner and Keegan 2001), in their organizational structure (Müller et al. 2014), in the running of projects (Winch 2001), in their risks strategies (Abednego and Ogunlana 2006), in the project process (Winch 2001), in project-based organizations (functional, matrix, or projectized) (Müller, Pemsel, and Shao 2014b), and in stakeholder management (Aaltonen and Sivonen 2009).

This leads to the second hypothesis:

- **Hypothesis 2.** The relationship between the project PMM and project success is moderated by project governance.
  - o **H2.1.** The impact of a comprehensive set of PMM elements on project success is moderated by project governance.
  - o **H2.2.** The impact of supplementing missing PMM elements on project success is moderated by project governance.
  - o **H2.3.** The impact of application of relevant PMM elements on project success is moderated by project governance.

4.4.4 Quantitative Research Model 4

Governance influences everyone indirectly through the governed supervisor and directly through subtle forces within the organization (Foucault 1980). Governance orientations vary considerably from an organizational perspective. The first orientation can range from extreme *shareholder* orientation, with a view to maximizing shareholder financial returns, to the other extreme of the continuum of a pure *stakeholder* orientation, in which stakeholder needs take priority over profit. The second orientation ranges from extreme *control* orientation of the continuum to extreme *behavior* orientation. Every organization's governance orientation can be located somewhere on the two continuums (Müller and Lecoeuvre 2014). Therefore, with governance being pervasive throughout an organization, coupled with the ability to gauge and also explain the consequences of the governance positioning using shareholder and stakeholder theory (from an organizational perspective) and agency theory and stewardship theory (from the perspective of behavior of individuals), it is interesting to understand and explain if and why project governance directly impacts project success. The operationalized governance paradigms from Müller and Lecoeuvre (2014) will be used to understand the relationship between project governance and project success. This then leads (see Figure 4.5) to the research question:

Does project governance have a positive impact on project success?

Section 3.3 describes the extensive literature on the link between corporate governance and corporate performance, which shows that weaker governance mechanisms have greater agency problems, resulting in lower corporate performance (Hart 1995; Hirschey, Kose, and Anil 2009; John and Senbet 1998; Ozkan 2007); greater shareholder rights have a positive impact on corporate

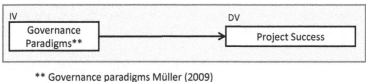

** Governance paradigms Müller (2009)

IV – Independent variable
DV – Dependent variable
MV – Moderator variable

Figure 4.5   Governance—Project Success Research Model

performance (Hirschey et al. 2009); and independent boards lead to higher corporate performance (Millstein and MacAvoy 1998). The link between governance and project performance (success) is implied, where project governance is seen as important in ensuring successful project delivery (Biesenthal and Wilden 2014).

This leads to the next hypothesis:

- **Hypothesis 3.** There is a correlation between project governance and project success.
  - o **H3.1.** There is a positive relationship between the governance orientation (shareholder–stakeholder) and project success.
  - o **H3.2.** There is a positive relationship between governance control (behavior–outcome) and project success.

4.4.5 Integrated Research Models 1, 2, 3, 4

Referring to Figure 4.6, the three studies (natural-science comparative, qualitative study, and the first quantitative study) are linked because of the use of PMM as an independent variable and project success as the dependent variable. In the qualitative study, the interviews did not restrict the scope environmental factors; whereas in the quantitative study, the focus was only on the environmental factor of project governance. Project governance was then divided into four paradigms based on Müller's governance paradigms (2009).

The natural-science comparative findings can be applied to both the qualitative and quantitative studies, and this is reflected in the discussion section of the overall findings in Section 6.5 (page 101).

The sequence of the studies started with the natural-science comparative to determine if a comparative could be built between the natural sciences and the social sciences (project management). Once the link was determined in the comparative (which was a PMM), the main part of the PhD was required to determine if the same phenomena could be observed and explained using traditional methods in social science (see Section 4.4.6 on philosophical triangulation). A mixed-methods approach was required, starting with the qualitative part to understand more about the concept of elements of a PMM and the impact of the environment (if any) on the relationship between the PMM elements and project success. Then the findings of the qualitative study and the literature review on project governance (see Section 3.3, page 23) refined the research model in terms of project context into the final research model. A fourth study using the quantitative data from the online survey was used to better understand the direct relationship between project governance and project success.

All four research models are shown in Figure 4.6.

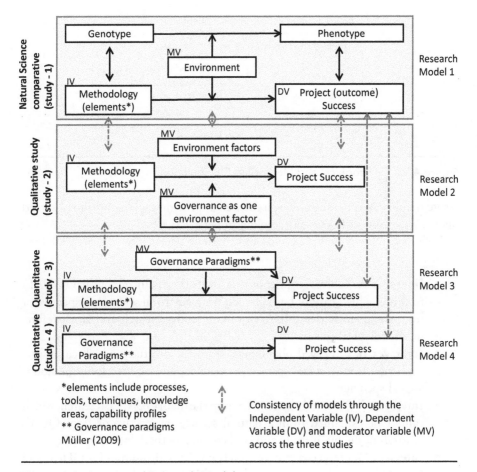

Figure 4.6   Integrated Research Model

4.4.6 *Philosophical Triangulation*

Philosophical triangulation is a way to potentially overcome predictable research results, because current research in project management is based on the same research methods (Tsoukas and Chia 2011). This research proposes using three epistemological perspectives (natural science based on positivism in the prestudy [Study 1], critical realism in the qualitative part of the sequential mixed methods [Study 2], and post positivism for the two quantitative Studies 3 and 4 [final part of the sequential mixed methods]). These alternative perspectives can be considered as a philosophical triangulation of results, which is defined by Denzin and Lincoln (2000) and Tashakkori and Teddlie (2009) as, "the combination of methodologies in the study of the same phenomenon." This should

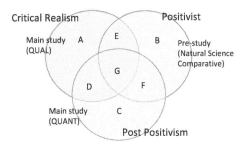

Figure 4.7   Philosophical and Methodical Triangulation

create stability in the research results as well as provide explanations to observed phenomena from different epistemological perspectives.

Referring to Figure 4.7, the philosophical triangulation intersects are explained below:

- **Intersect A.** Observed phenomena from the main qualitative (QUAL) research, which cannot be explained from a natural science perspective or from the quantitative (QUANT) main study research and is, therefore, methodological specific.
- **Intersect B.** Observed phenomena from the prestudy, which cannot be explained from a social science perspective, and is, therefore, methodological specific.
- **Intersect C.** Observed phenomena from the main QUANT research, which cannot be explained from a natural science perspective or from QUAL research of the main study and is, therefore, methodological specific.
- **Intersect D.** Observed phenomena from the main studies for QUAL and QUANT research providing a philosophical triangulation, but cannot be explained from a natural science perspective, and, therefore, this part (B) is methodological specific.
- **Intersect E.** Observed phenomena from the main QUAL study, which can also be explained from a natural science perspective providing a philosophical triangulation, but cannot be explained in the QUANT research, and, therefore, this part (C) is methodological specific.
- **Intersect F.** Observed phenomena from the main QUANT, which can be explained from a natural science perspective, providing a philosophical triangulation, but cannot be explained from the QUAL research, and, therefore, this part (A) is methodological specific.
- **Intersect G.** Observed phenomena, which can be explained from a natural science and both QUAL and QUANT social science perspectives, therefore providing a full philosophical triangulation.

This will be used in the discussion in Section 10.5 (page 201) to bring in an alternative perspective.

4.5 Prestudy (Study 1)

4.5.1 Data Collection Instrument Development

A literature review was conducted on genetics—or, more specifically, epigenetics, which is the study of stable alterations in gene expression potential that arise during development and cell proliferation (Jaenisch and Bird 2003)—and on project management, which is the application of knowledge, skills, and techniques to execute projects effectively and efficiently (PMI 2013a).

4.5.2 Validity and Reliability

There are a number of universal Darwinian extensions that apply the same criteria of eligibility as with Darwin's original evolutionary process criteria. They are:

- Variation in any given species
- Selection of the fittest variants—that is, those that are best suited to survive and reproduce in their given environment
- Heredity, where the features of the best-suited variants are retained and passed on to the next generation.

Darwin's evolutionary process criteria were applied to the concept of a PMM, including its elements, and all criteria were met. A number of scenarios were described using the attributes of both worlds to see if these were realistic and plausible, which they were. A theory-building exercise was carried out which first looked to find another comparative that was researched and accepted in the academic community and that exists in both the natural and social science worlds at two levels: a gene level and an organism level on one side, and an element level and the project outcome on the other. Complex adaptive systems (CAS) were selected as the theory-building comparative, because these systems exist in all four areas (natural to social sciences and at two levels), and CAS has become a major focus of interdisciplinary research in both social and natural sciences (Lansing 2003). The results of the theory building concluded that the attributes of CAS were present for each part of the natural-science comparative and, therefore, give greater validity to the comparative model.

Finally, the comparative was read by a professor of biology to determine whether the references on the natural science side were accurate, and the feedback supported this.

4.5.3 Limitations of the Research

Every research approach has its limitations. The comparative has used only mapping tables to compare attributes. Over the past 20 years, improvements to the comparative frameworks have been made in classifications and through the use of statistical methods to the degrees of relatedness in the comparative (Harvey and Pagel 1998; Martins and Garland 1991). The next step would be to look at statistical methods to see the degrees of relatedness in the comparative.

4.6 Main Study—Qualitative Research (Study 2)

4.6.1 Data Collection Instrument (Semistructured Interviews)

The empirical data collected in qualitative research provide richness based on real experiences with context that is not achieved from an online survey. However, rigor needs to be applied to ensure that data analysis techniques extract the most out of these data, as recommended by Miles and Hubermann (1994).

4.6.2 Sampling Approach

A theoretical sampling method was used to determine the list of interviewees, meaning the interviewees who have the best knowledge of the research subject. The number of interviews was determined by theoretical saturation, which means that when the answers from interviewees become convergent, and no new insights are gained for the concepts or categories, the sampling will stop (Miles and Huberman 1994). The data were collected from several industries and geographies so as to find commonalities and differences in order to understand the relationship between the variables.

4.6.3 Data Collection

A questionnaire was developed with six sets of questions covering PMM, project success, relationship of PMM to success, project environment, and other comments (see Appendix A to Chapter 6, page 107). The questions were derived from a literature review on the topics of PMM, success, and project context. The author conducted 19 semistructured interviews, then theoretical saturation was reached. Participants from 19 organizations in 11 business areas were categorized using the Reuters categorization system (Reuters 2013). The business areas included industrial, food and beverage, technology, financial, energy, telecom services, and research that spanned four countries (Switzerland, USA, UK, and

Germany). The level of the interviewees varied from project manager, program manager, and PMO lead to CTO and COO; therefore, some relevant information, especially regarding the usage of the PMMs and their purported strengths and weaknesses, needed to be considered based on the level of the interviewee.

The interviews were semistructured and lasted between 60 and 90 minutes. Interview notes and recordings were documented and compared for cross-validation. When additional questions or clarity were required on the responses, follow up was done using Skype sessions and email. This was needed for three of the interviews.

4.6.4 Data Analysis Method

Every interview was recorded, and notes were taken at the same time. Each interview was transcribed within a day. The transcripts were read several times in order to become familiar with the data in detail (Eisenhardt 1989) and then coded into the NVIVO software tool according to the topics and questions. Data reduction, data display, deducing conclusions, and verification techniques were used to analyze the interview data, following the guidelines outlined by Miles and Huberman (1994).

For example, the recorded data were broken down into concept units and provided with labels (codes). The first interviewee was asked to describe whether the PMM for any given project type integrates the "how to build" something with the "what to build," or is the "what to build" (requirement specifications) for the project type kept separately and why?

> . . . we have methods that are quite integrated in all that they deliver. So for example, in the CRM practice we have separate methodologies for Seibel and Salesforce.com. We are trying to get into manufacturing and the products that we would like to deliver are ones we would like to create. We would ultimately like to market end-to-end solutions, so how can they productize themselves using unique methodologies integrating the what and the how. The reasons why there are so many different types of in-house methods are because they are tailored to each area. The advantage of a methodology that integrates the what and the how is that it is a unique offering giving enhanced value and we should be able to corner a market. The disadvantage is that some clients may not want this (Miles and Huberman 1994).

By analyzing the meaning of the words "enhanced value," we see that a highly integrated PMM provides a competitive advantage in new markets and also aids in productizing an organization. So these words were coded to an integrated PMM and to a customized PMM.

As the interviews progressed, the same method was used to identify codes from the interviewees' responses. As new questions were asked, a PMM categorization was built to show the positioning of a PMM in terms of the origins and levels of customization and the benefits and disadvantages of integrating the "what" and the "how" into one PMM.

4.6.5 Validity and Reliability

Once the findings were derived from the interview notes and transcripts to ensure that the findings were credible, the checklist by Miles and Huberman (1994, pp. 278–279), which covers objectivity/confirmability, reliability/dependability/auditability, internal validity/credibility/authenticity, external validity/transferability/fittingness, and finally utilization/application/action/orientation, was used to inspect the analysis of the processes and the results. The reliability and validity are assured by considering the following:

Reliability

The data were collected from a spread of industries and countries (USA, Switzerland, Germany, and the UK). Identified patterns were cross-validated for reliability.

Validity

Concept validity was given through the theoretically derived model, which was built on existing literature, and from which the propositions were drawn. Construct validity was achieved through convergence of the interviewee data.

4.7 Main Study—Quantitative Research (Studies 3 & 4)

4.7.1 Data Collection Instrument

An online survey was used as the data collection instrument for the quantitative research.

4.7.2 Sampling Approach

A pilot test was performed to determine whether weaknesses existed in the design of the questionnaire. Ten respondents using a purposive sample were asked to carry out the survey and comment on the understandability of the questions, wording, logic, and length of time to complete. Based on the feedback, minor wording changings were made for clarity. The answers of the pilot were not used in the final analysis, because they were only used to improve the wording of the online survey.

Project, program, and project team members and functional managers were contacted using email with a link to the web survey. In addition, the survey details were placed on several LinkedIn forums for project management, which included the Project Management Institute (PMI), the International Project Management Association (IPMA®), and several other project management LinkedIn groups. An email with the survey link was sent to a number of PMI chapters. During April 2014, 386 respondents answered the survey within a period of 14 days. The following filter question was asked:

Do you have an understanding of your organization's or client's PMM where you have been involved as a project stakeholder, that is, someone working in or impacted by projects?

From the responses, 132 were disqualified through the filter question at the beginning of the survey and, therefore, were excluded from the survey. This resulted in 254 full responses that could be used for analysis. The respondents came from 41 countries, with 24% from Europe, 38% from North America, 22% from Australasia, and 16% from other countries. The average respondent's work experience was 22 years, and the average project-related work experience was 15 years.

An ANOVA test with a significance level of 0.05 was carried out between the demographic regions to see if there were differences in responses between the regions. The p value for the test was 0.249, showing no statistical differences between the regions. An ANOVA analysis was performed to assess if there was a difference between the mean project success rates for early and late respondents: The p value for the test was 0.149, showing no statistical differences in the means.

4.7.3 Data Collection

Data were collected through an online questionnaire. The questions on project success were developed by Khan, Turner, and Maqsood (2013). The questions on governance and the four paradigms within governance were developed by Müller and Lecoeuvre (2014). The questions on PMM were based on the qualitative study (Joslin and Müller 2016). Permission was granted by Khan, Turner, and Maqsood (2013) to use their scales. The online questionnaire is shown in Appendix A to Chapter 6 (page 107).

The questionnaire design is shown in Table 4.1, detailing the constructs for PMM, success, and governance, including the literature sources and scales used in the online questionnaire.

Table 4.1 Questionnaire Design—
PMM, Project Success, and Governance

Construct	Question	Source	Scale
PMM	Comprehensive set of methodology elements	Joslin and Müller (2015b)	5-point Likert scale, strongly disagree–strongly agree 5-point
	Supplemented missing methodology elements		
	Applied relevant methodology elements		
Project success	Project efficiency	Diallo and Thuillier (2004); Müller and Turner (2007b); Shenhar et al. (1997)	5-point Likert scale, strongly disagree–strongly agree 5-point
	Organizational benefits	Jessen and Andersen (2000); Thomas and Fernández (2008); Turner (2008)	
	Project impact	Bryde (2005); Diallo & Thuillier (2004); Müller and Turner (2007b); Wateridge (1998)	
	Future potential	Bryde (2005); Khang and Moe (2008)	
	Stakeholder satisfaction	Müller and Turner (2007b); Shenhar et al. (1997); Westerveld (2003)	
Project governance	Shareholder–stakeholder orientation	Müller and Lecoeuvre (2014)	Semantic differential scale 5-point Likert scale +2 to –2
	Behavior–outcome orientation		

The construct *project success* is based on five success factors: project efficiency, organizational benefits, project impact, future potential, and stakeholder satisfaction. The reliability of the five success factors is shown in Table 4.2, in which all factors have a Cronbach's alpha value over 0.7, indicating good reliability

Table 4.2 Reliability of Khan and Turner (2013) Success Factors

No	Factor	Items	% of Variance Explained	Crombach's Alpha
1	Project efficiency	8	15.9%	0.893
2	Organizational benefits	5	12.1%	0.796
3	Project impact	4	11.5%	0.811
4	Future potential	4	10.9%	0.762
5	Stakeholder satisfaction	4	10.5%	0.725
Total			60.9%	

(Field 2009). The construct *project governance* is based on 10 questions, where five are related to shareholder versus stakeholder orientation and the other five are related to behavior versus outcome control (Müller and Lecoeuvre 2014).

4.7.4 Data Analysis Method

Analysis was carried out following the guidelines from Hair et al. (2010). Data were checked for normality using skewness and kutosis measures of ±2. Boxplots of variables were done to identify outliers and t-tests between outlier respondents' answers and the wider sample to identify the representativeness of the answers of outliers. The answers from eight respondents appeared to be significantly different from the wider sample responses, and thus they were excluded from the analysis. This accounted for fewer than 3% of the valid responses and resulted in all variables meeting the thresholds for skewness and kurtosis, hence the data for the variables being normally distributed (Hair et al. 2010).

Exploratory factor analysis was carried out on the PMM, governance, and success variables to identify unknown, underlying structures and also to reduce the number of variables to a manageable size while retaining as much of the original information as possible (Field 2009).

Factor analysis was then used to determine the underlying dimensions for project context (governance) and project success characteristics. Following Sharma, Durand, and Gur-Arie (1981), hierarchical regression analysis was used to test the relationship between PMM and success (Hypothesis 1) and to test the moderating influence of governance on the relationship between PMM and success (Hypothesis 2). Finally, a number of ANOVA tests were used to compare the mean of groups including early and late responders, difference of geographical regions, difference of service- and product-based projects, project management experience levels, and comprehensiveness of methodologies to determine additional information pertaining to two or more of the research model variables.

4.7.5 Validity

Validity shows how well the concept is defined by the measures, whereas reliability shows the consistency of measures (Hair et al. 2010). Reliability of the data was carried out in different ways. Content validity was done by literature-based development of the measurement dimensions, and face validity was tested during the pilot. Construct validity was ensured through the use of earlier research results for the definition of the measurement dimensions, the development of the questionnaire (Joslin and Müller 2016; Khan et al. 2013; Müller and Lecoeuvre 2014), pilot testing of the questionnaire, and item-to-item and item-to-total correlations that were performed quantitatively through unrotated factor analyses. Testing item-to-item and item-to-total correlations showed that the required threshold values of 0.3 and 0.5, respectively, were reached (Hair et al. 2010).

Validity was tested through an unrotated factor analysis for each of the dimensions, which also served as the Haman test to exclude common method bias-related issues, as suggested by Podsakoff and Organ (1986). The results for governance, project success, and PMM factor analysis gave a Kaiser-Meyer-Olkin (KMO) sampling adequacy value of 0.8 or higher (with a significance of $p < 0.001$). KMO measures the intercorrelations between the variables through the measure of sampling adequacy (MSA). Kaiser (1974) recommends that acceptable values should be greater than 0.5, values between 0.5 and 0.7 are considered to be mediocre, values between 0.7 and 0.8 are considered to be good, values between 0.8 and 0.9 are great, and values above 0.9 are superb.

4.7.6 Reliability

Reliability was ensured by asking multiple questions per measurement dimension and testing for Cronbach's alpha values per measurement dimension being higher than 0.60 (Cronbach 1951). The Cronbach's alpha values for all of the measurement dimensions were greater than 0.747, which shows that the constructs are reliable (Hair et al. 2010)

The questionnaire was piloted by 10 respondents using a purposive sample. Only small wording changes were made to some of the questions to improve clarity. The pilot users' responses were not used in the final analysis.

This concludes the PMM chapter for studies 2, 3, and 4. The next chapter describes the objectives and aims of each study and provides a summary of the research findings.

Chapter 5

New Insights into Project Management Research: A Natural Sciences Comparative

Coauthored with Ralf Müller
BI Norwegian Business School, Norway

5.1 Introduction

In this chapter, a new research perspective toward project management phenomena is developed; it builds on the existing natural science theory of genotyping and phenotyping by developing a contemporary comparative model for project management research, which uses natural science molecular biology (genomics) as a way to investigate social science (specifically, project management) phenomena. The comparative maps concepts and terminology and, in doing so, explains why phenomena in genomics (study of genetics) can be compared with practices, behaviors, and established thinking in project management. To support the theory-building process, the attributes of complex adaptive systems (CAS) are used to validate the constructs of the research. The comparative is then used to answer the research question by identifying two social science phenomena—"lessons intentionally not learned" and "bricolage of competing methodology subelements"—followed by a detailed explanation of the reasons for the phenomena using the attributes of the comparative. This article provides further examples of phenomena that were derived from the comparative model as well as the types of research questions for which the model would provide insight.

The authors believe that using a comparative model will challenge established thinking so that many aspects of project management will be seen in a new light in both the research and practitioner communities of project management.

Over the past 40 years, project management research has grown and matured. The methods and techniques used today provide well-established frameworks for designing and executing research studies. However, the success of these established approaches had some unforeseen consequences, because research questions are often limited by the methodological starting positions and possibilities (Williams and Vogt 2011). Research designs determine the nature of the results; therefore, a limited set of research methods will impact the variance of research designs, which in turn leads to almost predictable results. Drouin, Müller, and Sankaran (2013) succinctly described this design dilemma by stating: "If we always do what we always did then we should not be surprised that we always find what we always found."

Contemporary methods have been developed and applied in many fields of scientific activities; these methods have provided for the development of new theories that challenge established theories and provide for fresh and alternative explanations of phenomena (e.g., Alvesson and Sköldberg 2009; Flyvbjerg 2001). The purpose of this article is to suggest that context-related concepts of natural science can be used as a theoretical lens for research in projects and their management—for example, in social phenomena such as projects. The concept of genotyping and phenotyping is used to exemplify the use of natural science perspectives to social science phenomena. Underlying this concept is an objective ontology applied to real entities (reifying a project as a "thing"), using the epistemological stance of process and/or variance methods in the sense of Van de Ven and Pool (2005).

The aim of this study is to contribute to transformative research by suggesting a particular empirical natural science perspective for some social science phenomena, such as research in project management methodologies and project portfolios as a comparative to existing perspectives. The National Science Foundation (NFS) (2007) describes transformative research as involving:

> *Ideas, discoveries, or tools that radically change our understanding of an important existing scientific or engineering concept or educational practice or leads to the creation of a new paradigm or field of science, engineering, or education. Such research challenges current understanding or provides pathways to new frontiers.*

To achieve this purpose, the following research question is posed: How can a natural science perspective be used to understand social science phenomena?

The results of the research will benefit both the academic and practitioner communities by providing alternative perspectives on project management, which should provide new insights into project management phenomena from a different epistemological perspective. In addition, the ability to create stability in the research results through the combination of methodologies in the study of the same phenomenon will be provided.

The next section contains a literature review of the theories behind a comparative analysis and then covers complex adaptive systems, which is the perspective taken for theory building. The comparative model is then described and is followed by two examples of how the model is applied. The article concludes with a discussion and conclusion.

Please note that all natural science terms used in this chapter are defined in Table 5.1 (beginning on next page).

5.2 Literature Review

5.2.1 Comparatives

One of the most powerful tools used in intellectual enquiry is comparison, because any observation made repeatedly gives more credence than a single observation (Peterson 2005). Boddewyn (1965) describes comparative approaches as those concerned with the systematic detection, identification, classification, measurement, and interpretation of similarities and differences among phenomena. The disciplines such as social science, including project management, usually rely on observation rather than experimentation, unlike the natural sciences, in which randomized experiments are the ideal approach for testing hypotheses. However, some research problems cannot readily be addressed using experiments—for example, when looking at research involving two or more species in evolution, ecology, and behavior (Freckleton 2009).

Comparative approaches have been used for decades to address the limitations of experiments, in which virtually every field in biological sciences uses comparatives (Gittleman and Luh 1992). Comparative analyses, unlike experimental studies, have historically relied on the simple correlation of traits across species. Over the past 20 years, improvements to the comparative frameworks have been made in classifications and the use of statistical methods to the degree of relatedness in the comparative (Harvey and Pagel 1998; Martins and Garland 1991).

Comparatives have been made between natural and social sciences using metaphors, such as the book *Images of Organization* (Morgan 1997) and biological comparatives—for example, cells of an organism with organizational knowledge (Miles et al. 1997) and behavioral characteristics of a group of

(text continues on page 58)

Table 5.1 Terms and Definitions

Term	Definition	Source
Altruism	Disinterested and selfless concern for the well-being of others.	*Oxford Dictionary*
Cell	In biology, the smallest structural and functional unit of an organism, which is typically microscopic and consists of cytoplasm and a nucleus enclosed in a membrane.	*Oxford Dictionary*
Chromosome	A thread-like structure of nucleic acids and protein found in the nucleus of most living cells, carrying genetic information in the form of genes.	*Oxford Dictionary*
DNA	Deoxyribonucleic acid, a self-replicating material which is present in nearly all living organisms as the main constituent of chromosomes. It is the carrier of genetic information.	*Oxford Dictionary*
Evolutionary stable strategy (ESS)	In game theory, behavioral ecology, and evolutionary psychology, an evolutionarily stable strategy (ESS) is a strategy which explains why altruism is not sustainable.	John Maynard Smith
Fidelity	The degree of exactness with which something is copied or reproduced.	*Oxford Dictionary*
Fitness landscape	In evolutionary biology, fitness landscapes or adaptive landscapes are used to visualize the relationship between genotypes (or phenotypes) and reproductive success.	Sewall Green Wright
Gene	(*Informal use*) A unit of heredity which is transferred from parent to offspring and is held to determine some characteristic of the offspring. (*Technical use*) A distinct sequence of nucleotides forming part of a chromosome, the order of which determines the order of monomers in a polypeptide or nucleic acid molecule which a cell (or virus) may synthesize.	*Oxford Dictionary*
Genome	The entirety of an organism's hereditary information.	Dawkins (1974)
Genotype	The genetic constitution of an individual organism. Often contrasted with *phenotype*.	*Oxford Dictionary*

(continues on next page)

Table 5.1 Terms and Definitions (cont.)

Term	Definition	Source
Heredity	The passing on of physical or mental characteristics genetically from one generation to another. The relative influence of heredity and environment.	*Oxford Dictionary*
Lineage	A sequence of species, each of which is considered to have evolved from its predecessor: e.g., the chimpanzee and gorilla lineages. A sequence of cells in the body which developed from a common ancestral cell: e.g., the myeloid lineage.	*Oxford Dictionary*
Methodology	A system of practices, techniques, procedures, and rules used by those who work in a discipline	Project Management Institute
Methodology element and subelement	A part of the methodology. A methodology element may contain one or more subelements, which are then termed *units of knowledge*.	Authors
Mimicry	In evolutionary biology, the close external resemblance of an animal or plant (or part of one) to another animal, plant, or inanimate object.	*Oxford Dictionary*
Nucleus	A dense organelle present in most eukaryotic cells, typically a single rounded structure bounded by a double membrane, containing the genetic material.	*Oxford Dictionary*
Organism	An individual animal, plant, or single-celled life form.	*Oxford Dictionary*
Phenotype	An organism's phenotype is its observable characteristics or traits, both physical and behavioral.	Malcom and Goodship (2001)
Pleiotropic	The production by a single gene of two or more apparently unrelated effects. Pleiotropy occurs when one gene influences multiple phenotypic traits.	*Oxford Dictionary*
Progenotype	Progenotype is used to denote the project core makeup (project methodology and the methodology elements).	Authors

(continues on next page)

Table 5.1 Terms and Definitions (cont.)

Term	Definition	Source
Project	A temporary endeavor undertaken to create a product, service, or result.	PMI *PMBoK® Guide*, 6th Ed.
Project element	An essential or characteristic part of a project.	Authors
Project outcome	The results of a project in terms of deliverables and non-deliverables, irrespective of whether the original project success criteria were achieved.	Authors
Traits	A genetically determined characteristic.	*Oxford Dictionary*
Unit of knowledge	The smallest unit of information that is able to take on the state of being true of false.	Authors

organisms known as complex adaptive systems, or with organizational leadership (Schneider and Somers 2006). Few have gone beyond the juxtaposition but still have provided new insights into explaining phenomena that may not have been discovered or explained without the comparative.

Discussions about the appropriateness of the natural or social science approaches to research in projects and their management often refer to the context independence of natural science research. A frequently drawn conclusion is that all social phenomena (such as projects) are context dependent, and therefore, natural science research approaches are deemed inappropriate for gaining an understanding of social phenomena (e.g., Flyvbjerg 2001). This perspective may be appropriate in some research studies but presents an oversimplification in others. A great deal of natural science research takes place in context-dependent situations, and just as much social science research takes place in situations of contextual independence. For example, Knorr-Cetina (1981, p. 358), who analyzed the differences in research situations between natural and social science, concluded:

> . . . that the situational logic of natural and technological science research appears similar to the situational dynamics inherent in social method, and that this similarity is strengthened by the apparent universality of interpretation in both social and natural science method. Given this similarity, it is time to reconsider customary routine distinctions between the social and the natural science which ascribe to the former what they deny to the latter. And given this similarity, it may be time

to reconsider scientific method in general as just another version of, and part of, social life.

In the field of project management research, comparatives are made mainly through theoretical lenses, including complexity theory, agency theory, contingency theory, and complex adaptive system theory (Eisenhardt 1989a; Hanisch and Wald 2012; Holland 1992). Some of these theories, such as complexity theory and complex adaptive system theory, are derived from observing nature (Brown and Eisenhardt 1997; Holland 2012). Comparatives are done between two things of interest that may not have been researched—for example, project managers and career models (Bredin and Söderlund 2013).

From the literature, it is clear that there is a need and a benefit in using comparative approaches in the field of project management. A great deal of the man-made world is based on nature and its evolutionary principles, including humans gaining insights by comparing species or comparing a part of an organism (such as a cell or a gene) with the phenotype and behavioral characteristics of that organism.

Dawkins (1988) stated that, "Biology is the study of complicated things that give the appearance of having been designed for a purpose." Project management can be inherently complex in terms of achieving desired and designed outcomes within volatile environments. There are many similarities between biology and project management in terms of complexity, design, impact of changing environments, and product lineage.

There is a literature gap in comparing the core makeup and characteristics of an organism with the core makeup and characteristics of project management.

Creating a new comparative, like any other type of analysis, requires that the phenomena are compared and abstracted from a complex reality. Therefore, it is important to provide a focused, careful delineation of scope, use of defined and accepted terms, and development of assumptions (Boddewyn 1965). The focus and delineation of scope, including use of terms for the natural-science comparative model, are described in Section 5.3.

5.2.2 Complex Adaptive Systems

A section on complex adaptive systems has been included in the literature review because this concept is used to build a theoretical validation of a comparative model.

The study of complex adaptive systems—a subset of nonlinear dynamical systems—has become a major focus of interdisciplinary research in both the social and natural sciences (Lansing 2003). To understand the concepts behind complex adaptive systems, it is important to note that complex adaptive systems have a large number of components, often called *agents,* that interact and adapt or learn (Holland 2012).

Nonlinear systems are ubiquitous, as mathematician Stanislaw Ulam observed (Campbell et al. 1985); they exist in the worlds of both natural and social sciences, in which examples in the natural world include ant colonies, swarms of bees, flocks of birds, cells, and the nervous system (Rammel, Stagl, and Wilfing 2007). In the man-made world, examples include the internet, power grids, cites, and societies (Holland 1992).

Complex adaptive systems were first derived from systems theory and cybernetics in the 1950s (Ashby 1957; Carnap, Fechner, and Hartmann 2000), then in the 1960s the term *complexity science* took hold, and from there *complex adaptive systems* evolved. Since the late 1980s, complex adaptive systems have been used to model virtually every aspect of our world, including impacts of disruptions in weather, earthquakes, communications, transportation, energy, and financial systems, as well as to influence management practices and project management research (Shan and Yang 2008).

Because complex adaptive systems exist in both the natural and social science worlds and are well researched, these make an obvious choice for building the theory behind the natural- to-social-science comparative model that is described in the next section.

5.3 Introducing the Comparative Model

The natural- to-social-science comparative model was developed with the aim of determining whether observations to phenomena in the natural science world could help to provide alternative explanations to social science phenomena. To be able to develop the model, a decision had to be made as to which of the sciences was most applicable to social sciences. Project management can be inherently complex in terms of achieving desired and designed outcomes in volatile environments. Biology was a natural choice, but physics was also a candidate. There are similarities between biology and project management in terms of complexity, design, impact of a changing environment on biology and project management, lineage, and heritage. Biology was selected because it is the study of complicated things that give the appearance of having been designed for a purpose, whereas physics is the study of simple things that do not tempt us to invoke design (Dawkins 1988).

Referring to Figure 5.1, the comparative model, which is also the research model, shows the linkage between the natural science and the social science worlds. Starting with the natural science part of the model, a genotype (Greek *genos,* "race" + Latin *typus,* "type") is the genetic makeup of a cell or an organism.

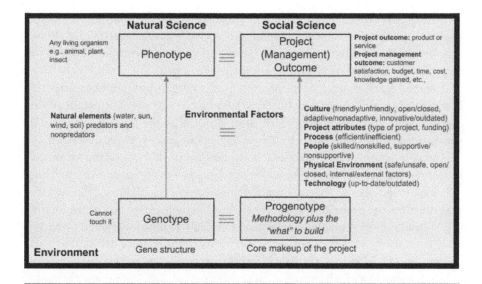

Figure 5.1   Comparative Model and Its Attributes (Research Model)

A phenotype (from Greek *phainein*, "to show" + *typos*, "type") is the composite of an organism's observable characteristics or traits—that is, something physical, which in most cases can be seen by the human eye. Every organism from the point of conception is influenced by the environmental conditions that have a direct impact on its phenotype. The term *phenotype* results from the expression of an organism's genes, the influence of environmental factors, and the interactions between the two. A genotype is the genetic makeup of a cell, containing the information of what and how to replicate in order to ultimately create the organism and keep it alive (Boulding 1978).

Moving to the social science part of the model, the term *progenotype* denotes the project's core make-up included in the lived project methodology, its elements (which are the parts of the methodology), and the requirements of what to build. The progenotype includes all the information needed to create the project outcome (product or service) and, ultimately, the information needed to maintain and enhance the product or service.

A project's environment is described in terms of what impacts the progenotype (i.e., the project core make-up) and how the environment impacts the development of the project. In natural science, the equivalent is the particular environment impacting a genotype of an organism. Using this comparison, there is a way to compare the genotype with the progenotype when the environmental factors impact both worlds (genotype and progenotype), resulting in genes and methodology elements being used (switched on) or not used (switched off). The switching effect on genes and elements throughout the life of an organism and the respective project life cycle results in observable characteristics in a phenotype or project outcome, called *traits*. These traits can then be traced back to the respective genes/elements in the genotype/progenotype.

The model also shows that the phenotype of an organism (the organism itself) is comparable with the project outcome (product or service). The details of this aspect of the comparison are described in depth after the topic of evolution of organisms and evolving project methodologies (progenotype) are covered.

The comparative model provides no indication of how organisms or project outcomes (products or services) evolved or adapted to the environment over time, which is an underlying factor in the development of the comparative. The next section describes the evolutionary aspects of the underlying principles of the model.

Over billions of years, organisms have evolved by constant gradual evolution from bacteria to what they are today (Dawkins 1988). Project deliverables such as cars, buses, cities, and all their infrastructure and subcomponents have also evolved, but over a much shorter evolutionary period. Darwin's theory of evolution (Darwin 1859) states that organisms that have successfully evolved are the best-suited variants optimized for their environments, and these organisms in turn create offspring, which then start a new round of evolution. This reproductive cycle can be viewed as an evolutionary algorithm that creates and/or forms the fitness landscape for the organisms that are best adapted to the then-given environment (Wright 1932).

According to Darwin (1859), the evolutionary process has three components:

- Variation in any given species
- Selection of the fittest variants—that is, those that are best suited to survive and reproduce in their given environment
- Heredity, where the features of the best-suited variants are retained and passed on to the next generation

The social science concept of evolution is similar to that of natural science, in that product or service evolution is within a project environment. The meaning of evolution within the social sciences, specifically project management, is the new release of a product/service (project outcome) in which the procedures for problem solving and trial and error indicate an evolutionary process at work. More specifically, product/service evolution refers to searching for the best solution for any given problem on how to meet success criteria by entering trials, testing performance, eliminating failures, and retaining the successes. This all assumes that the environmental conditions change within known boundaries; otherwise, there is a risk of project outcomes (product/service) becoming obsolete.

The same is true for the natural science world, in which examples of species unable to cope with the drastic changes to their environment include dinosaurs, the dodo, and the Irish elk. Today, many species are in imminent risk of extinction due, in part, to their inability to adapt quickly enough to their changing habitats.

In the social science world, products can be designed to adapt according to the environment within a given range, but when the environment changes too much, there is a likelihood of obsolescence. The advantages over nature are that a replacement product can be designed for current and future environmental conditions, with or without the design lineage (genes) of the predecessor product.

One key difference between natural and social science is that humans can predict, to some extent, the impending environmental changes. This is achieved by applying intelligence and tools/techniques to the problem. Decisions can be made to obsolete a product or continue with a product's evolution (lineage). In the natural science world, an organism does not have the ability to prepare itself in a noncyclical changing environment, as there is no foresight. An organism either adapts or becomes extinct.

In summary, mutations in organisms are random, but evolution is not. Evolution promotes the survival of species through natural selection. Product/service evolution is structured through reasoning, with the underlying premise of being competitive—that is, "fit" for purpose.

5.4 Characteristics of a Natural-Science Perspective

In this section, three specific characteristics were used to build the comparative model and should be kept in mind when applying the suggested perspective to social science. These are complexity, replicator, and Universal Darwinism.

5.4.1 Complexity

Evolution has no boundaries, irrespective of whether it is in the field of natural or social science. Jean-Baptiste Lamarck believed the evolution of organisms was a one-way road, which he called *complexity force* or, in French, *le pouvoir de la vie* (Lamarck, 1838). In social science, the management and development of products or services within project and programs are also becoming more complex (Vidal, Marle, and Bocquet 2011).

Complexity is a regular topic in senior management circles within and outside of the project environment (Hitt 1998). Many project influencers talk about reducing project complexity. This statement is easy to make without understanding the complexities and challenges to achieving a successful project outcome. If the complexity discussion were to be moved to the natural science field to build an organism, it is unlikely that the same comment on complexity would result. The concern from the project influencers is really about *unnecessary* complexity, and not complexity itself. Evolution in both natural and social sciences is resulting in greater complexity (Adami, Ofria, and Collier 2000; McShea 1991) but should not be overly complex—one could suggest a sort of practical application of Occam's razor, which means, "When there are two competing theories that make exactly the same predictions, the simpler one is the better" (Thorburn 1918).

5.4.2 Replicator

The goals of every organism are to survive and replicate so its genes have the greatest chance of survival over generations (Dawkins 1974). The term *replicator* was first developed by Darwin (Darwin 1859) in natural science. So what is the equivalent of the term *replicator* in the project world? If a product or service is going to be successful, then the progenotype must be resilient and have a high fidelity at the element level to ensure that it always creates a successful outcome. To achieve this, the progenotype and the project outcome need to be replicated as many times as possible to build up a base for justification of future product updates. This, in turn, will help determine whether the product starts and/or continues with a lineage or not.

5.4.3 Universal Darwinism

Darwin's theories have been generalized over the years, and all generalizations fall under the grouping *Universal Darwinism* (Dennett 1996). To date, two categories exist within Universal Darwinism: gene-based and nongene-based extensions. Gene-based extensions cover areas including physiology, sociology, and linguistics, whereas nongene-based extensions cover areas including complex adaptive systems, memetics, cultural selection, and robotics.

This chapter is based on a gene-based Universal Darwinism extension to derive the model and a gene- and nongene-based Universal Darwinism extension (called complex adaptive systems) to build the theory behind the model.

With these characteristics in mind, we can now develop the comparative model.

5.4.4 The Comparative Model

The comparative model in Figure 5.2 shows the basis for a two-level comparative (Levels 1 and 2) between natural science and social science.

The genotypes (genes) are the starting points because, through their expression, they will impact the organism's phenotype, but not the other way around—a one-way causation. The first comparative (Level 1) is between the genotype

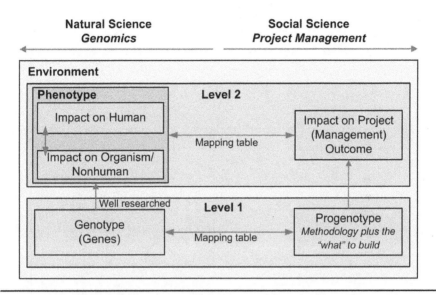

Figure 5.2   Two-Level Comparative Model

and the methodology elements (progenotype). The second comparative (Level 2) compares an organism's phenotype with a project outcome (phenotype).

Figure 5.3 (on next page) provides a detailed two-level mapping table between genotype and progenotype and organism and project outcome. The bottom half of the table shows where the key attributes of an organism's genes have been described and compared with the key attributes of a project's progenotype.

Every organism has a unique genome, which contains, through encoded DNA, the entirety of the organism's hereditary information. The genome describes what unique organism and how to build it. Comparatively, a procedure-based methodology that has been updated through lessons learned derived from a product with lineage (versions) contains all the information on how and what product to build. Genes have enduring attributes that have ensured their survival over millions of years. A progenotype also has enduring attributes that will determine if it will survive over the course of time or be replaced with something more adaptable to the environment.

Referring to the Level 1 comparative in Figure 5.3, there is a striking similarity between the attributes of a gene and the attributes of an element of a progenotype. For example, in gene backup versus safeguarding of project knowledge, both reduce the risks of losing unique knowledge. New gene creation from existing genes versus distribution of project knowledge ensures that the genetic/project procedures are at the right place and time—creating new genes

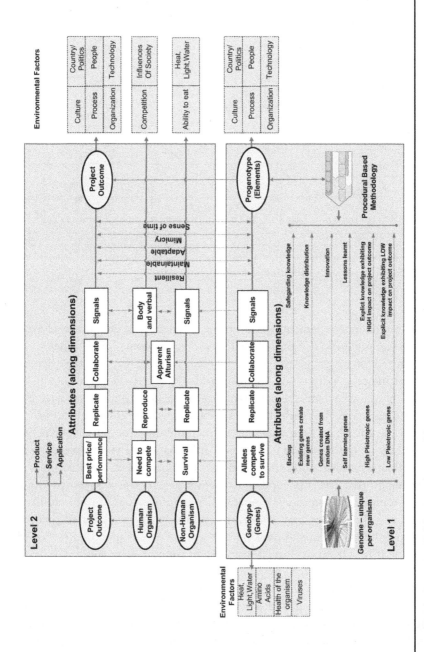

Figure 5.3 Mapping Table: Levels 1 and 2

from DNA versus project innovation (i.e., the ability to *create*) and, finally, self-learning versus project lessons learned (i.e., the ability to *adapt*).

How are genes, versus the elements in a progenotype, controlled to ensure that the described attributes are realized? There are several types of genes—one of them is the "master gene" (Pearson, Lemons, and McGinnis 2005). A master gene controls and monitors the progress of the other genes within its domain. The control of genes is totally decentralized. Comparatively, the elements of a progenotype are controlled with something equivalent to a master gene called *local governance.* Learning from the study of genomics, it would make sense to control progenotypes by decentralized updates like the master gene concept. Progenotypes, like genomes, contain a vast amount of context-related information. A single person is unlikely to have the knowledge to decide which content needs to be updated according to context. If a person attempts to update a progenotype without expert knowledge in the specific field, it will likely lead to a suboptimal result, which in turn would reflect in the project performance and ultimately impact the project outcome.

Wikipedia is built on the concept of decentralized updates using experts in their knowledge domain. One person invariably takes the lead as a subject matter expert coordinating other contributors. This is similar to the master gene concept in natural science. If this decentralized approach were taken to update a progenotype, then topic experts would also decide which progenotype's elements would be the most appropriate for each project's profile including context.

In the natural science world, natural selection at the gene level is where a competing gene—that is, a gene that has two or more alleles (or competitors)—vies for selection and becomes the dominant gene, and the nonselected genes become recessive (Mendel, 1865). However, in future generations, it is possible that recessive genes could be selected on the basis of environmental and non-environmental reasons. Recessive genes can cause problems in the organism, which may or may not be seen in the organism's traits (Dewey et al. 1965). It is also possible to select an element within a progenotype, which is not as applicable as its alleles (equivalent approaches); this may cause problems, which may or may not be observable as a project trait. There are certain genes that greatly impact their phenotype's traits, which are called *high-pleiotropic genes,* and other genes that have less impact on their phenotype's traits (*low-pleiotropic genes*). The same is true for the elements within a progenotype; some elements will have a higher impact on the project outcome (product/service) than others.

Now we move on to the Level 2 comparison of an organism's phenotype and the project outcome (product/service). The key attributes of an organism are mapped to the key attributes of a project outcome. The path back to the gene (genotype) is shown to ensure a consistency of comparison.

The majority of a gene's attributes (except for mimicry and signals) is directly related to the gene's ability to survive and replicate (Wickler 1968). The mimicry and signal attributes (described later) have been included because they indirectly help the genes to survive by means of the organism.

A gene competes against its alleles to determine the dominant gene (Gagneur, Elze, and Tresch 2011). Nonhuman organisms compete for survival; however, humans compete for paid jobs and battle to stay in their jobs, which often leads to territorial behavior. A project outcome, typically a product or service, competes against other similar products or services in terms of price, quality, and performance.

Replication is a prerequisite for survival in both the natural and social science worlds. When a gene replicates, its fidelity is one in 100 million (Pray 2008), where fidelity means the degree of exactness with which something is copied or reproduced. Environmental conditions may cause mutations in cells during the replication process. Influences such as radiation, chemicals, pollution, and viruses can all impact an organism's cells and, therefore, the DNA/genes contained within (Lewtas et al. 1997. Products are also replicated with degrees of fidelity. The quality control checks ensure that the replication process stays within predefined tolerances. Products, like organisms, suffer from defects that may be undetected by the quality control checks but are likely to be observed during the products' lifespans.

Collaboration

The term *collaborate* is used in the context of genes and project outcomes, whereas *apparent altruism* is used for organisms. Genes that don't compete (nonalleles) collaborate to produce phenotypic effects to support their organism's survival (Nelson 2006). This could be in the form of signaling or other similar traits. The project outcome (product/service) is often designed to collaborate with other products and/or services—for example, other component parts, internet services, servers and infrastructure, or software. Whenever there is an interface from one product or service to another, it is a form of collaboration. Collaboration is normally associated with organisms, but there is no reason why products and services cannot be considered to collaborate by interfacing to support their collective needs within any given environment. Organisms (human and nonhuman) collaborate where there is mutual benefit, but they appear also to do altruistic things, acting with disinterested and selfless concern for the well-being of others.

This raises the question of why altruistic actions exist if there is no personal benefit. In evolutionary biology, altruism contradicts the theory of natural selection (Dawkins 1974). There are many explanations concerning altruism within nonhuman species, and all of them point to an underlying self-interest. A mathematical model using game theory was created by Maynard Smith (1982) called *evolutionarily stable strategy* (ESS), which shows that altruism does not pay off in the survival of a species. A similar model called the prisoner's dilemma, also using game theory, shows why two individuals might not cooperate, even when it appears that it is in their best interests to do so (Nowak and Sigmund 1993). Humans have more complex motives than animals, but the underlying acts of altruism always include aspects of self-interest for both humans and animals (Fehr and Fischbacher 2003; Simon 1993). Collaboration in humans with apparent altruism is really just collaboration in which both parties will benefit.

Signaling

Signaling is a phenotype trait created by gene expression that helps the organism to survive (Wickler 1968). Signaling is the conscious act of switching on and off something that warns or attracts a recipient of the signal. Products and services have built-in signally systems for both attracting and warning the recipients of the signals.

Resilience

Resilience is a feature that genes have built up by using various techniques described in the Level 1 mapping. To some degree, organisms and humans are resilient to environmental conditions; accordingly, a product or service also needs to be resilient to environmental conditions.

Maintainability

When a gene or an organism cannot maintain itself, it will die. Likewise, when a product or service is not maintainable, it will fall into disrepair and soon be replaced with something that is more maintainable.

Adaptability

When a gene or organism, or likewise a product or service, cannot adapt to the environmental conditions, it will most likely become extinct or obsolete. Some organisms have learned to become adaptable, but only when the change to the environment is not too extreme and/or when the change does not occur too quickly (Williams, 2008).

The same is true for a product or service where environmental conditions could render it obsolete when the designed degree of adaptability is not sufficient to function.

Mimicry

Mimicry is a phenotype trait that is created by gene expressions that help the organism to survive by mimicking other species (Wickler 1968). The same trait occurs in the product and service world when better-known branded products and services are mimicked to increase the likelihood of survival of the mimickers.

Sense of Time

Organisms exhibit a sense of time using a biological process called a *circadian rhythm* (Yerushalmi and Green 2009). This rhythm, which oscillates in 24-hour cycles, is widely observed in plants, animals, fungi, and cyanobacteria. Products and services are also time cognizant to ensure that maintenance and upgrade windows do not overlap with operational times. Projects that create products and services also work to time through their schedules to ensure deadlines are met.

5.5 Theory Building

Many scholars define theory in terms of relationships between independent and dependent variables. Other scholars have defined theory in terms of narratives and accounts (DiMaggio 1995). According to Eisenhardt, theory is evaluated primarily by the richness of its account, the degree to which it provides a close fit to empirical data, and the degree to which it results in novel insights (Eisenhardt 1989b). As the constructs of the comparative model are new, no empirical data exist to substantiate or disprove the model. In the absence of any previous comprehensive theory building, this section aims to builds theory by using the established constructs of complex adaptive systems and, in doing so, will provide validity for the constructs of the natural- to-social-science comparative.

There are two ways to use complex adaptive systems for the comparative model theory building. Referring to Figure 5.4, the first way, which is the simplest, is

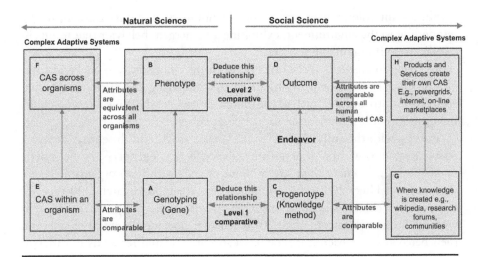

Figure 5.4 Theory Building Using Complex Adaptive Systems in Both Natural and Social Sciences

to describe the attributes of complex adaptive systems and then show how these exist in both the natural and social science worlds for each entity. The second way is to review the literature that describes complex adaptive systems comparatives for each entity pair—for example, complex adaptive systems within an organism and genotype. This chapter focuses on the first approach by comparing the attributes for each entity across both worlds for Levels 1 and 2.

Complex adaptive systems are special cases of complex systems—that is, dynamic networks of interactions and relationships (Holland 2006). These cases exist in natural sciences and social sciences (Miles et al. 1997) and within and across organisms (Holland 1992). As such, these complex systems exhibit Darwinian properties of variance, selection, and heredity (Hodgson and Knudsen 2006). Examples of organisms that are part of complex adaptive systems include ant colonies, swarms of bees, flocks of birds, and humans in societies (Rammel et al. 2007). Examples of complex adaptive systems within organisms include cells, the nervous system, and the immune system (Holland 1992). In the man-made world, examples of complex adaptive systems include the internet, power grids, cities, and societies (virtual and physical).

Using Holland's definition of complex adaptive system attributes, agents within a complex adaptive system are self-similar and numerous, hence are seen as complex. The agents' behavior within the complex environment anticipates responses and therefore exhibits emergent behavior, allowing complex adaptive system self-organization (Holland 2012). An example of an agent is an ant within an ant colony, hence numerous in numbers. The ants are continually responding to the environment, exhibiting an emergent behavior that results in a collective self-organizing process.

Referring to Figure 5.4, the links between the comparative model and complex adaptive systems will be discussed by comparing the attributes of each entity. If the attributes of each of the linked entities are the same or comparable, then this provides a foundation for the comparative.

Starting with the entity "genotyping" labeled as (A) and the entity "complex adaptive systems within an organism" labeled as (E), a genotype is the genetic makeup of a cell where there are millions of cells within an organism (Feder, Bennett, and Huey 2000). In a human, there are approximately 200 types of cells, and all exhibit similar characteristics, with more than one trillion cells within a human (Bianconi et al. 2013). There is no central control of cells and their behaviors, but each cell type has a specific role and responds to different situations. Cells exhibit the characteristics of complex adaptive systems and so have been labeled as such (Lansing 2003). Therefore, the genotype that exists within all cells, within the nucleus, in the DNA, chromosomes, and genes, are considered a complex adaptive system (Holland 2002).

The second link is between the "phenotype" labeled as (B) and the "complex adaptive systems across organisms" labeled as (F). A phenotype is the expression of its genes (genotype) and as such is a living organism (Jaenisch and Bird 2003). Organisms that live in large groups such as ants and humans are similar in themselves, exhibit emergent behavior, and self-adapt to the environments they live in (Holland 1992). These attributes are consistent for any organism that lives in large groups (Dawkins 1974). The conclusion can be drawn that a phenotype (organism) that is part of a large community is also part of a complex adaptive system.

The third link is between the "progenotype," that is, project methodology and its elements labeled as (C); and a wiki, which is a complex adaptive system, and labeled as (G). A methodology can contain thousands of elements called *units of knowledge* (Joslin and Müller 2013). The elements of any methodology are typically created by one or more individuals who have combined and recombined processes that are based on knowledge spanning over 100 years. For example, *A Guide to the Project Management Body of Knowledge® (PMBOK® Guide)*, Fifth Edition (Project Management Institute 2013) had more than 250 contributors and reviewers working on producing the body of knowledge, for which most of the source material can be traced back through the previous editions to the originators of the units of knowledge. The origin of the Office of Government Commerce (OGC) PRINCE2® methodology (OGC 2002) was based on a predecessor called Prompt II.[1] Many Prompt II elements were

---

[1] Prompt II was developed by Simpact Systems Ltd in 1975.

derived from third-party concepts—for example, Gantt charts, the Program Evaluation and Review Technique (PERT), and procurement.

To remain relevant, every methodology needs to evolve within the context of the environment for which it was designed. This is achieved by managing the methodology elements in terms of creating new elements to add new knowledge and removing or changing one or more methodology elements to update existing knowledge. A methodology contains many methodology elements and subelements, among which the subelements are considered to be at the lowest level.

Methodologies can and do evolve in a similar manner as wikis. A wiki is defined as a website or database developed collaboratively by a community of users, allowing any user to add and edit content (Pearsall, Soanes, and Stevenson 2011). Wikipedia, the most well-known of all wikis, is updated in a decentralized and uncoordinated way by approximately 100,000 individuals.[1] Wikis exhibit all of the attributes of a complex adaptive system (Andrus 2005; Nikolic and Davis 2012), and therefore methodologies that are being constantly evolved and adapted in a wiki-like environments[2] are also complex adaptive systems.

The fourth and final links are the *project outcome* labeled as (D)—that is, a product or service; and *complex adaptive systems across products or services* in the social science world, labeled as (H). Man-made complex adaptive systems such as power grids, the internet, automated driverless cars, autonomous robots, and online marketplaces are derived from products and services that are used or configured to coexist in an environment with other similar or identical products (Shenhar and Bonen 1997). Products that are similar, used in large numbers in an evolving, connected way with no central control, exhibit the characteristics of complex adaptive systems (Holland 2006).

In summary, the four constructs of the model exhibit complex adaptive system attributes, in that they are either complex adaptive systems in themselves or one of many agents described as a complex adaptive system. Therefore, the comparative of a genotype labeled as (A) to a progenotype labeled as (C) is demonstrated both from the comparative mapping described in this article and by the comparison of complex adaptive system attributes previously discussed. The same logic applies to the relationship between a phenotype labeled as (B) and the project outcome labeled as (D).

---

[1] http://www.quora.com/

[2] Harvard University and Cornell University are two universities with wiki-based project management methodologies. Other project management wikis can be found by searching on wiki + project management methodology.

5.6 Application of the Comparative

Based on the discussions thus far, a natural science perspective suggests the genotype as the independent variable and the phenotype as the dependent variable, with the environment as the moderator variable.

To simplify the explanation, the following environmental factors (i.e., moderator variables) are described in a project and natural science (genotype) perspective as follows:

- Individual (personality and traits of a project manager)
- Organization (culture)
- External environment that the organization is in (stable or volatile)

The independent variable progenotype is subdivided into elements and subelements, wherein each subelement can be considered to be a unit of knowledge, which is analogous to a gene being a unit of heredity.

There is no formal definition of a unit of knowledge within the field of project management. Therefore, the following working definition is used for this article: A unit of knowledge is the smallest unit of information that is able to take on the state of being true or false. Using this definition, a methodology subelement can now be defined as a unit of knowledge constituting an affirmation being the smallest unit that can be true or false.

A progenotype (i.e., parts of a methodology in its environment) is applied to a project to achieve a desired outcome. A progenotype describes what knowledge is required and how this knowledge is to be applied to a project in order to achieve a successful project outcome.

A project contains processes, tools and techniques, deliverables, and stakeholders, which can be referred to as *project elements*. The sum of the elements constitutes a project. A project element in this context is defined as an essential or characteristic part of a project.

Assuming project outcome traits are measurable, we can state the following hypotheses:

- **H1.** There is a direct relationship between progenotype (project methodology) and project outcome.
  - o **H1a.** The relationship between progenotype and project outcome is moderated by the project environment.

A unit of analysis is the relationship between the progenotype and the project outcome.

When applying the model, the following example is used:

A project manager has experience that was gained from several project implementations. Some of the projects created new versions of the product. The lessons learned from previous versions of Product 1 were fed back into Project 1's unique progenotype. Some of this knowledge was generalized and put into the organization's generic progenotype.

Both the generic progenotype and the unique Project 1 progenotype are evolving like an organism's genome when it replicates. Both progenotypes are adapting through the lessons learned, which benefits the next generation of projects. However, the generic progenotype does not have a lineage (unlike the unique Project 1 progenotype). In an evolutionary sense, when an organism's genome is always based on an average mix of genes within a species, it cannot evolve (Darwin, 1859) and probably is extinct after an epoch. With this comparative, there is a risk that when an organization tries to use a generic progenotype over a period of time without some level of customization (e.g., to a product, an organization type, or a project type), the projects that use the generic progenotype are suboptimally run. Organizations that endorse the generic progenotype approach are likely to become uncompetitive against companies that have customized the projects' progenotypes and reaped the full benefits in the outcomes of the resulting projects. The conclusion is that the generic progenotype will no longer be used, and in the worst case, organizations adopting a generic progenotype approach that don't adapt it will eventually go out of business.

Continuing with the project manager example, the same project has a newly assigned project manager. The project progenotype has evolved with every new product release. In the applied research model example, the project manager is considered to be an environmental factor and can decide whether to implement Project 1's unique progenotype or change the units of knowledge of the progenotype by:

- Replacing them
- Leaving some out
- Complementing existing units of knowledge with his or her own personal units of knowledge

The changes to the unique progenotype may or may not improve the traits of the project during the project's development (embryonic stages) and in the project outcome. If the units of knowledge are excluded and are not replaced with something equivalent, there is a high probability that the deficiencies in the methodology will appear as project traits during the project development, project outcome (product or service), and project management outcome. Project management outcome traits would include increases to cost, scope impact, delays in schedule, and decreased customer satisfaction.

The project manager now decides to substitute units of knowledge from the unique progenotype with his or her own units.

Depending on how these units of knowledge are integrated into the progeno-type and how applicable the units are to the project environment, the project traits will be influenced either in a positive, neutral, or negative way. There is an equivalent in natural science wherein the genome of a species has been modified by a virus or another organism (larva) that splices (changes) the DNA structure by introducing its own genes (Dawkins 2004). The effect is that the change in the phenotype and behavior of the organism during its embryonic and fully grown stages are mainly to the benefit of the larva or virus and less so to the organism itself (Dawkins 2004). The comparative is where the project manager changes the unique progenotype's "genome" to achieve the project outcome but may also personally benefit from the changes. This would not have been the case if the progenotype had been implemented without change.

Two natural science examples are given, one with a negative outcome and the other with a positive outcome:

- **Negative outcome.** The introduction of a virus that creates havoc in an organism or a gene mutation that results in a hereditary disease, which often results in a premature death.
- **Positive outcome.** A gene evolution, giving phenotype traits that provide an advantage over the species that don't have this mutation.

Very few gene mutations result in positive outcomes—most result in nega-tive outcomes (Loewe 2008). Could this be a word of warning for the project managers who are considering the alteration of established progenotypes, who don't have an in-depth understanding of the project environment, or who don't know how the units of knowledge in the progenotype interact with each other within that project environment?

For project managers with little or no experience who venture to change a highly evolved progenotype (derived from a product with lineage), there is a likelihood of a failed project outcome (if it ever gets to that point). In the natu-ral science world, some viruses cause havoc in the infiltrated organism, and the result is that the organism's immune system is triggered, which normally kills the virus after a hard fight. The analogy in the social science world is the "inexperienced project manager" who vastly deviates from a highly evolved progenotype without understanding the implications that can trigger the organization's immune system. The trigger is the organization's "governance" and results in a similar outcome—removal of the project manager, but prob-ably not before harm is done to the project in terms of wasted resources and damaged reputation.

There are two other environmental factors described in the model, which would also act as moderator variables: organization culture and external market environment. Depending on the culture and the state of the external market environment (stable or volatile), both can either positively or negatively impact the project during the embryonic stages and final project outcome.

The progenotype contains many units of knowledge that relate to different parts (elements or subelements) of a project—for example, financial, planning, scheduling, or risk aspects. When a project manager leaves out one or more units of knowledge from the progenotype, the resulting project traits should be traceable back to the cause of the problem. However, when all the environmental factors impact the project in some way—for example, through inexperienced actions of project managers, closed environments, volatile markets, and so forth—this will impact multiple project traits and will make it difficult to determine which project traits are symptoms and which are root causes.

The determination of the root cause(s) may be further complicated, because each unit of the progenotype will have varying degrees of impact on the project traits (called *pleiotropic effect*). Projects that are out of control are often misdiagnosed when symptoms are addressed and root cause(s) are ignored. This happens as the result of a lack of understanding of the cascading cause-and-effect issues in complex environments. In the project world and in the natural science world, most issues can be traced back to a maximum of one or two root-cause issues. The challenge is to quickly find them before there is irreparable damage to the project.

Until now, the project manager has been described as an environmental factor, wherein the unit of analysis is the impact of the progenotype on the project outcome. However, the project manager is also an organism driven by his or her own genes' need for survival. This gives rise to a second level of comparison: Level 2 in Figure 5.3, between the human (organism) and the project outcome. The unit of analysis now becomes the impact of the project manager (and his or her team) on the project outcome. With both levels (1 and 2) in the comparative, the real world of project management is more accurately modeled; however, the downside is added complexity in applying the comparative. Two example questions are posed here, the answers to which are derived by the Level 2 mapping shown in Figure 5.3:

- The lessons-learned feedback loop is an important part of ensuring a progenotype (methodology) evolves. However, why does it seem that lessons sometimes are intentionally not learned?
- Are project progenotypes (methodologies) complementary, or are they really competing with a bricolage of individual units of knowledge through use and copy across progenotypes?

These project management–related questions are discussed using the proposed perspective.

5.6.1 Lessons Intentionally Not Learned

An example of lessons intentionally not learned is when a project manager believes he or she knows better and makes a decision not to use part(s) of the unique progenotype (project methodology) that have evolved over several project generations. This is a conscious decision not to learn or use knowledge gained from his or her predecessors. The question is, why does a project manager believe he or she knows better, when clearly a great deal of knowledge and experience has been synthetized from project learning into a continually improving progenotype?

One explanation taken from the natural science perspective is that organisms are driven by survival instincts. A human (in this case, the project manager) strives to survive in the world he or she knows and will use all available resources that are believed to provide him or her with the maximum advantage. Taking something that has been developed by someone or a group of people does not necessarily provide an advantage, nor does it differentiate, because the project manager is genetically driven to succeed by competing in the same environment. Humans have intelligence and the ability to understand the implications of risks. However, achieving success in a workplace (irrespective of how success is measured) often overrides the implications of the risk events, especially when the environment is new and the risks are not fully understood. Lessons not learned in projects do not lead to a fatality, unlike in the animal world, in which this would inevitably lead to a fatal mistake. If the implications were the same in the project world, then every lesson would be learned based on the assumption that the project manager is capable of assimilating and integrating the new knowledge.

5.6.2 Bricolage of Competing Methodology Subelements (Units of Knowledge)

Every gene fights for survival with its allele(s), and so does every methodology subelement (unit of knowledge). Looking at the individual genes within an organism's genome, each gene's goal is to replicate and be present in as many organisms as possible within that species (Dawkins 1974). The same is true for every unit of knowledge within the progenotype. Once a unit of knowledge is selected for any given project, it no longer needs to compete and therefore will collaborate with all other units of knowledge within the progenotype to increase the probability of a successful project outcome. However, regardless of whether or not a unit of knowledge is selected for any given project, its goal is to be used (replicated) in as many projects' progenotypes as possible. This will create a bricolage of individual units through use and copy across progenotypes. Will the individual units of knowledge survive the course of time? It will depend on the success of each project and, therefore, the combination of units of knowledge for each project environment. Only the most aligned progenotypes for any given project environment will survive.

5.7 Discussion

Comparatives have been made over the past decades between natural and social sciences, providing new ways to view and compare the items being compared, but few comparatives have gone past the juxtaposition. Developing a contemporary method to observe a phenomenon can be ridiculed, but the findings in using the new approach soon offset the skepticism (Kuhn 1970).

The two-level comparative that is based on well-defined terminology, set assumptions, and detailed mapping tables goes further than many comparatives. This is because the apparently separate disciplines do have many similar characteristics in terms of complexity, design, impact of changing environments, and lineage. The underlying concepts of the comparative (Universal Darwinism, evolutionary stable strategy, phenotyping/genotyping) allow for a rich comparative that can be extended to encompass existing concepts within natural sciences, such as eusocial organisms (Kramer and Schaible 2013).

Complex adaptive systems—an area of great interest in the academic community that has also been well researched since the 1980s—has helped to provide the theory building and support for the constructs of the comparative.

Not only can the comparative be used to provide alternative insights into project management research questions, but it also can be used to identify phenomena.

Table 5.2 provides examples of phenomena that were identified using the comparative, including the level at which they are relevant within the comparative.

The following are examples of research questions that can provide an idea of what can be addressed by the comparative:

- Are independent or lone projects at greater risk of being canceled in a project portfolio and, if so, why?
- Is a generic or customized methodology more likely to achieve the project goals and, if so, why?
- With all the best practices, why are lessons not learned, and what can be done about it?
- How can we better educate senior management about project management, and what role can lessons learned play?
- What is the range of project durations that have the best chance for achieving the project goals and why?

Table 5.2 Examples of Phenomena That Were Identified Using the Comparative

Phenomena	Comparative Framework— Level (1 or 2)	Description
Selfish projects	Level 2	Projects compete for resources such as management time, funding, and skills; therefore, no interest to work with other projects unless a mutual interest exists.
Lessons not learned	Level 2	Lessons are intentionally not learned, as project managers cannot differentiate themselves or prove their intellectual ability (fitness).
Methodologies with bricolage of competing elements	Level 1	Methodologies elements are competing to be selected for a project to provide the best fit for the context of the project and environment. Once selected, they collectively work together to deliver a well-designed project outcome adapted to the environment.
Impotent (generic) methodologies	Level 1	Not designed for any project type or environment. When competing against other methodologies that are adapted, it will have reduced chance of its own survival and that of its project outcome.
Lesson learned fighting for management attention	Level 2	Lessons learned are reified with the objective of being consumed by management, so they are learned. Takes the perspective of the lessons needed to achieve their objectives.
Naturally aging projects	Level 2	Projects age, and in doing so become less effective. Understanding the attributes of aging and implications of project efficiency and effectiveness, as well as traits in the project outcome, will provide insight on how to set up and structure projects.

- Are customized project methodologies more appropriate for projects that deliver products?
- What is the impact of a rapidly changing environment on a project's methodology effectiveness?
- Are the performances of foreign (nonlocal) project managers better or worse than local project managers (in what areas), and does their performance substantially improve when compared with local project managers over time (adaption)?

Any research question that has the potential to be answered in part from an evolutionary, altruistic, or methodological perspective could benefit from using the comparative or an extension of it.

5.8 Conclusion

This study shows how the comparatives within and across disciplines are able to answer questions as well as provide new insights that may not be possible with existing research techniques. A great deal can be learned by modeling and mapping the natural science world to bring new perspectives on topics that conform to natural order.

The two-level comparative model developed in this study shows how a natural science perspective can be used in understanding social science phenomena and how well-established research areas such as complex adaptive systems can be used in the theory building to support a new comparative.

The strengths of the findings show that using a new perspective through the lens of natural science is likely to bring insights that challenge conventional thinking in project management. The weaknesses in the findings are that all comparisons must go beyond the juxtaposition of phenomena that are potentially comparable, requiring further research into the explicit contrasts and explanations. Also, the use and extension of this model require proficient knowledge in both sciences, plus an understanding of the implications of the mapping as well as in identifying and mapping new attributes.

The authors believe that most project management themes could be applied to the model, which will provide new and interesting observations in project management research.

5.8.1 Future Research

This study contributes to transformative research by suggesting a particular empirical natural science perspective for social science phenomena, such as research in project methodologies, reifying projects, and project outcomes as a comparative to existing perspectives. This study should help project practitioners take a new perspective on how they view projects in terms of progenotype and how the elements of the progenotype are assembled to create a project outcome most suited to its environment. Any deviation from the perspective or its implementation will result in a suboptimal project and project outcome performance.

Future research recommendations are:

- To apply the comparative model in existing research areas in order to understand how it performs in terms of supporting current findings, challenging current findings, and discovering new findings.

- To understand the limitations and strengths of the comparative model.
- To extend the comparative model along the attribute dimensions to allow a broader scope of applicability. For example, the attribute dimension *collaborate* can be extended to include social organisms (Danforth 2002; Simon 1960), which will provide insights into understanding why independent projects in a project portfolio may be at a greater risk of being canceled or put on hold than linked or related projects.

Chapter 6

The Impact of Project Methodologies on Project Success in Different Project Environments

Coauthored with Ralf Müller
BI Norwegian Business School, Norway

6.1 Introduction

Project failures are estimated to cost hundreds of billions of euros yearly (McManus and Wood-Harper 2008) and are not limited to any specific region or industry (Flyvbjerg, Bruzelius, and Rothengatter 2003; Nichols, Sharma, and Spires 2011; Pinto and Mantel 1990).

Project methodologies have been developed specifically to help address low success rates using project-related knowledge (The Standish Group 2010; Wysocki 2006). Government bodies have helped to establish standards in methodologies and guidelines, with their tools, techniques, processes, and procedures (Morris et al. 2006). The term *project methodology* implies a homogeneous entity; however, is it a heterogeneous collection of practices that vary from organization to organization (Harrington et al. 2012). To understand the impact of the relationship between methodology and success, the building blocks of a methodology need to be understood. These building blocks are not defined or agreed upon to an extent that they are commonly accepted; therefore, we define the building blocks of a methodology as methodology elements that can include processes, tools, techniques, methods, capability profiles, and knowledge areas.

The reference to processes within the above definition is not to be confused with the project life cycle. A process is defined as a structured set of activities to accomplish a specific objective (TSO 2009), whereas a project life cycle is defined as the series of phases that a project passes through from its initiation to its closure (PMI 2018).

The literature on project methodologies is divided. There is a positive attitude toward project methodologies, and sometimes unrealistic expectations are directed toward them (Lehtonen and Martinsuo 2005). However, when these methodologies do not produce the expected outcomes, they are replaced by other methodologies—often with those that have other limitations (White and Fortune 2002). The two main topics in research on project methodologies are linked with whether project methodologies should be *standardized* (Breese 2012; Milosevic, Inman, and Ozbay 2001; Milosevic and Patanakul 2005) or *customized to the project environment* (Lechler and Geraldi 2013; Payne and Turner 1999; Pinto and Mantel 1990). Research has shown that projects in which methodologies are used provide more predictable and higher success rates (Lehtonen and Martinsuo 2006; Wells 2012). However, there are still high project failure rates for projects that do use project methodologies (Wells 2012).

More research is required to better understand how project methodologies impact success, but it would be naïve to assume that phenomena occur without the influence of context. This is also implied in the literature—for example, there is much research to determine whether standardized or customized project methodologies lead to greater project success.

Governance influences organizations, in that it "provides the structure through which the objectives of the organization are set" (OECD 2004). Governance influences people indirectly through the governed supervisor and directly through subtle forces in the organization (and society) in which they live and work (Foucault 1980). Governance in the area of projects takes place at different levels at which there is project governance on individual projects—namely, "the use of systems, structures of authority, and processes to allocate resources and coordinate or control activity in a project" (Pinto 2014, p. 383). As governance influences organizations, as well as multiple aspects of project management, it is also likely to influence the value created by project management, especially the effectiveness of a project methodology and its impact on project success.

This study uses project governance as the context variable.

The purpose of this study is to investigate whether there is a relationship between a project methodology, including its elements, and project success, and if this relationship is impacted by the project environment (e.g., project governance or culture). This will provide the knowledge for organizations to customize project methodologies to their environment, thereby minimizing the risk of methodology elements being used suboptimally while also allowing "at-risk" methodology elements to be proactively monitored.

To achieve the study's purpose, the following research question is posed:

What is the nature of the relationship between the project methodology, including its elements, and project success, and is this relationship influenced by the project environment, notably project governance?

The unit of analysis is the relationship between project methodology and project success.

The overall methodological approach of the study is inductive. The authors qualitatively validate the research model (see Figure 6.1) through interviews that are inductively analyzed.

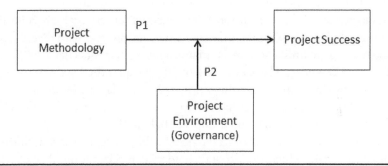

Figure 6.1   Research Model

Results from this research should qualitatively validate the constructs of a theoretically derived research model by clarifying terminology to gain insights for a future study on methodologies, their elements, and their impact on project success.

The next section provides a literature review of the research subject, followed by a description of the methodology in this study, an analysis section, a discussion, and conclusions. Appendix A to this chapter (page 107), provides the interview questions and analysis data.

6.2 Literature Review

This section reviews the literature on project methodologies, the possible moderating effect of the project environment on the relationship between methodology and success, and the definition and measure of project success.

6.2.1 Project Success

The classification of a project as a success or a failure is, to a degree, subjective (Ika 2009). Müller and Jugdev (2012) describe project success as "predominately in the eyes of beholder," meaning one stakeholder may consider a project successful, whereas another stakeholder would consider it a failure. To reduce the subjectivity relating to project success, a common understanding is required. To achieve this, success criteria should be defined in the initiating phase of the project (PMI 2018, p. 51). Morris and Hough (1987) define success criteria as the measures used to judge the success or failure of a project; these are dependent variables that measure success.

It is worth mentioning that even with comprehensive definitions for project success criteria, some project criteria remain subjective by nature—for example, product usability or the acceptance of new processes. The methods and techniques aimed at quantifying subjective measures reduce subjectivity. However, when subjective criteria are mixed with objective criteria, which collectively determine whether a project is considered a success, projects with diverse groups of stakeholders are unlikely to reach unanimous agreements (Ika 2009).

Project success criteria have evolved from simple quantifiable time, scope, and cost measures (Iron Triangle), which primarily are related to project efficiency (Bryde 2005), to measures that have a longer-term perspective directly relating to effectiveness and organizational impact (Belout 1998; Jugdev, Thomas, and Delisle 2001; Shenhar, Levy, and Dvir 1997). Project success is a multidimensional construct in which project stakeholders can select a number of project success criteria they believe are important by which to judge success.

For each project, not only should success criteria be defined from the beginning of the project, but also, the relevant success factors need to be identified and incorporated in a timely manner across the project life cycle (Pinto and Prescott 1988).

Neither PRINCE2® nor *A Guide to the Project Management Body of Knowledge® (PMBOK® Guide)*, 6th Edition (PMI 2018), define the term *success factors,* but both standards make use of the term. Turner (2007) defines project success factors as elements of a project, which when influenced increase the likelihood of success; these are the independent variables that make success more likely.

These definitions for project success factors and project success criteria will be used in the interviews as well as in the quantitative research to ensure a common understanding of terminology.

The selection process for determining relevant success factors is not without risk. When success factors that have absolutely no impact on the project outcome are implemented, both management time and cost is wasted (Atkinson 1999). The selection and/or timing of the implementation for nonrelevant success factors are called Type 2 errors (Atkinson 1999). Type 1 errors are success factors that are important but incorrectly implemented. Attention should be given not only to the selection of individual success factors but also to the combination or grouping of related success factors that are contingent on the project life cycle (Belassi and Tukel 1996). To understand the complex interaction of success factors throughout the project life cycle, success factor frameworks were developed (Belassi and Tukel 1996). A framework is defined as a basic structure underlying a system or context (Pearsall, Soanes, and Stevenson 2011). Therefore, a success framework can be defined as a basic structure, underlying system, or context that supports the project life cycle to meet the project's success criteria.

In the area of project management research, success frameworks typically consist of concepts, definitions, and existing theory for a particular study. Some of the success frameworks described in the literature relate to success *criteria,* others to success *factors* (Ika 2009). In both cases, success frameworks can vary from being conceptual, with a list of success factors or success dimensions (where the latter is associated with success criteria), to more practitioner-oriented, in which figures illustrate lists and/or groups of success factors that may have process flows or links relating to project life cycles. The use of success frameworks should help to reduce Type 1 and Type 2 errors but must be selected according to the context of the project (Shenhar et al. 2002).

Project success is the dependent variable.

6.2.2 Project Methodologies

During the past 40 years, attention has shifted from individual tools and methods to methodologies that encompass multiple methods and tools (Lehtonen and Martinsuo 2005). However, the transition to methodologies has created inconsistencies in how the terms *method* and *methodology* are sometimes used. For example, PRINCE2®, which is a process-oriented project methodology, is described as "a method that supports some aspects of project management" (TSO 2009), and PMI's body of knowledge is often referred to by practitioners as a project methodology, which academics point out is a body of knowledge. Anderson and Merna (2003) have helped to categorize the methodologies into process models, knowledge models, practice models, and baseline models.

Merriam-Webster (2013) defines a method as "a systematic procedure, technique, or mode of inquiry employed by or proper to a particular discipline." A methodology comprises many methods, wherein each method is applied in a particular situation. Therefore, a methodology is considered to be the sum of all methods and the related understanding of them. The term *project methodology* implies a homogeneous entity; conversely, it is a heterogeneous collection of practices that vary from organization to organization (Harrington et al. 2012). To understand the impact of the relationship between methodology and success, the building blocks of a methodology need to be understood. The authors describe the building blocks of a methodology as *methodology elements* that can include processes, tools, techniques, methods, capability profiles, and knowledge areas. These methodology elements can then be applied to a project, as needed, throughout the project life cycle.

Research on project methodologies is limited, and the results are somewhat contradictory. For example, literature is split on whether project methodologies directly contribute to goals (Cooke-Davies 2002; Fortune and White 2006; White and Fortune 2002) or to the perceived appropriateness of project management (Lehtonen and Martinsuo 2006). Another example is that, in some cases, the existence of positive attitudes toward project methodologies, and, in other cases, unrealistic expectations are directed toward project methodologies (Lehtonen and Martinsuo 2005). However, if these methodologies do not produce the expected results, they are replaced by other methodologies, often with methodologies with other limitations (White and Fortune 2002).

A third example is a critical attitude toward methodologies because they sometimes do not seem to fit, for example, complex project environments; however, when methodologies are customized, they tend to be too complex to be maintained, and the organization may switch from an overly formal, rigid control to chaotic freedom (Lehtonen and Martinsuo 2005). Thomas and Mullaly (2007) explain this dilemma by citing "the multiplicity of potential benefits that executives, practitioners, and consultants associate with implementing project management [methodologies], but they make no effort to quantify these values . . . where empirical evidence exists it is tantalizingly fragmented and incomplete." Perhaps this problem is a result of something that lies deeper in the elements of a methodology. Busby and Hughes (2004) have an interesting notion that methodologies are being infected with pathogens, especially in the tools and systems employed that impact project success. This implies that, irrespective of configuration, when the tools and systems used in a methodology are infected with pathogens, the methodology never achieves its intended purpose of supporting project success.

Methodologies are referenced in the literature either as a whole (The Standish Group 2010) or by one or more aspects of project management practice methodology element(s) and investigating the impact of these practices on project success (Cooke-Davies and Arzymanow 2003; Cooke-Davies 2002; Milosevic and Patanakul 2005). To understand how methodologies and their elements collectively support achieving project success, viewing methodologies at too high a level or on a singular element basis may not be sufficient. Guidance may come from looking at project success factors that are described at the level of the methodology elements. The difference between a methodology element and a success factor is in the description. A success factor contains an adjective used to describe its syntactic role to qualify the underlying methodology element. For example, project scheduling is a methodology element, whereas *efficient* project scheduling is a success factor.

Taking one methodology element at a time and determining its impact on project success does not give a holistic picture of how the elements of a methodology impact the characteristics of the project success. Some methodology elements may have a greater collective impact on project success characteristics than others.

There is a gap in the research regarding whether the elements within any given methodology collectively impact the characteristics of project success.

- **Proposition 1.** There is a positive relationship between project methodology elements and the characteristics of project success.

6.2.3 Project Environment's Moderating Effect on Project Methodology and Project Success

The Standish Group placed the selection and use of a project methodology as one of the top 10 factors contributing to project failure (The Standish Group 2010). The report states that project methodologies have provided improvement to project success (35%), in contrast to the rate of project failure (19%) and projects that partly met their project success criteria using project methodologies (46%). The conclusion is that closer attention should be given to the correct choice and application of the methodology and tools. Cooper (2007) observed that many organizations are mismanaging projects because they are using tools and techniques that are not appropriate for the project type or applying financial selection criteria that are not appropriate for the project type.

Lehtonen and Martinsuo (2006) sum up the research dilemma on project methodologies by stating, "The confusion in research results is reflected also in companies' swing between standardized and customized systems, and between formal and chaotic methodologies." A conclusion can be drawn from the literature that the effective use of a methodology is contingent upon the project environment. This statement may at first appear contrary to the term *standardized methodologies,* but it is unclear from the literature as to the origins of implemented standardized methodologies. Regardless of whether a standardized methodology is derived from an international standard or alternatively developed in-house, both examples suggest degrees of customization, even though they are classified in the literature as standardized methodologies. Therefore, to fully understand whether a methodology is standardized or customized, the origin of the methodology needs to be understood.

The extensive research on success factors topics, such as leadership competency profiles (Müller and Turner 2010), stakeholder management (Turner and Müller 2004), risks addressed (Cooke-Davies 2002), realistic schedule (Morris and Hough 1987), and HR management (Belout and Gauvreau 2004), all take into consideration the project context, which may or may not be reflected and/or used within the respective organizations' project methodologies.

There is a research gap regarding the impact of the project environment on the relationship between an applied project methodology and its elements on project success.

- **Proposition 2.** There is a moderating effect of the project environment, notably governance, on the relationship between a project methodology and project success.

In this study, project governance is considered part of the project environment.

6.2.4 Contingency Theory and the Theoretical Perspective

To support achieving the aims of this research, an appropriate theoretical lens is contingency theory. Contingency theory, which was first developed over 50 years ago, suggests that there is no single best way to manage and structure an organization (Burns and Stalker 1961; Woodward, Dawson, and Wedderburn 1965). Contingency theory has since been applied to project context research, with the first studies in the late 1980s (Donaldson 2006).

The application of contingency theory in the field of project management has been applied to various areas, including topology of projects with minor and major impacts (Blake 1978), innovation types in business (Steele 1975), product development project types (Wheelwright and Clark 1992), leadership styles for project and functional managers in organization change (Turner, Müller, and Dulewicz 2009), project procedures customized to context (Payne and Turner 1999), leadership styles per project type (Müller and Turner 2007), and project type and the ability to select appropriate management methods linked to project success (Boehm and Turner 2004; Shenhar and Dvir 1996). Contingency theory will be used to help explain observed phenomena relating to the influence of environmental factors, notably project governance, on the relationship between project methodology and project success.

The literature implies the relationships shown in Figure 6.1 but does not indicate that these relationships have been tested. The literature review also indicates a lack of understanding about the relationship between methodology elements and their impact on success characteristics and the possible moderation by the project environment, notably project governance.

6.3 Research Methodology

A philosophical stance of critical realism was used in the study. Critical realism assumes that reality is mostly objective; however, social constructions are recognized, which must be outlined in a subjectivist way (Alvesson 2009). This paradigm combines people's subjective interpretations, framed by their experiences and their view of reality, with objective mechanisms and events (Bhaskar 1975).

A deductive approach was taken to validate the model shown in Figure 6.1. Data collection was done through semistructured interviews. Interviews were used to gain a greater depth of understanding as to how the interviewees understood the way in which project methodologies performed within their environments in terms of impacting the characteristics of project success and whether the project environment influenced the relationship of project methodology and project success. Project methodologies are described using different terminologies; therefore, a definition was required to create a generic understanding of the parts of a methodology. The findings will be used for a follow-up larger study to achieve generalizable results.

6.3.1 Development of Data Collection Instrument

The interview questions were derived using contingency theory as a theoretical lens (see the Appendix to this chapter on page 107).

Six sets of questions were addressed:

- Nature of the organization and the type of projects run within the organization
- Project methodology(s); how it was originally developed and evolved, project types supported, strengths, and weaknesses
- Project success; organization definition
- Impact of a project methodology (including its elements) on project success
- Impact of the project environment (including project governance) on the relationship between methodology and the characteristics of project success
- Other comments from the interviewees rating project methodology(s), project environment, and project success

The first set of questions was used to obtain an understanding about the organization's business area, core business, and size and types of projects, including complexity, technical challenge, and pace. The questions relating to project types and characteristics (urgency, complexity, and technology) were taken from Shenhar and Dvir (2007) and (TSO 2009) and are included in Table 6.1. These questions should provide some context regarding the choice of the organization's methodology(s) and level of customization.

Table 6.1 Interview Data Overview

Chapter	Comments	Journal/Conference Proceedings
Chapter 5: New Insights into Project Management Research: A Natural Sciences Comparative	Natural- to social-science comparative including theory-building section and comparative section	EURAM 14th Annual Conference, Valencia, Spain, June 2014 Joslin, R., & Müller, R. (2015a). New insights into project management research: A natural sciences comparative. *Project Management Journal*, 46(2), 73–89.
Chapter 6: The Impact of Project Methodologies on Project Success in Different Project Environments	Qualitative part of the PhD	PMI Research Conference, Portland, Oregon, July 2014 Joslin, R., & Müller, R. (2016b). The impact of project methodologies on project success in different project environments. *International Journal of Managing Projects in Business*, 9(2), 364–388.
Chapter 7: Relationships Between a Project Management Methodology and Success in Different Project Governance Contexts	Quantitative part of the PhD	Joslin, R., & Müller, R. (2015b). Relationships between a project management methodology and project success in different project governance contexts. *IJPM*, 33(6), 1377–1392.
Chapter 8: The Relationship Between Project Governance and Project Success	Quantitative research based on data obtained from the online survey	Joslin, R., & Müller, R. (2016c). The relationship between project governance and project success. *IJPM*, 34(4), 613–626.
Chapter 9: Using Philosophical and Methodological Triangulation to Identify Interesting Phenomena	Qualitative part of the PhD	Joslin, R., & Müller, R. (2016d). Identifying interesting project phenomena using philosophical and methodological triangulation. *IJPM*, 34(6), 1043–1056.

The second set of questions relates to the methodology(s) within the organization, in order to understand whether methodology is based on an international or internally developed standard, whether there are variations of the methodology for different project types, and what are its strengths and weaknesses.

The third set of questions concerns the definition and interpretation of project success—that is, whether project success criteria were defined within the organization, and whether there is any written data.

The fourth set of questions addresses the impact of project methodology and its elements on project success.

The fifth set of questions refers to the moderating effect of project environmental factors on methodology and project success and then focuses on one moderating environmental factor—project governance.

6.3.2 Sampling

Convenience sampling was used to determine the interviewees' list, meaning the interviewees who have the best knowledge of the research subject. The number of interviews was determined by theoretical saturation (Miles and Huberman 1994). The data were collected from several industries and geographies so as to find commonalities and differences in order to understand the relationship between the variables (see Figure 6.1).

6.3.3 Data Collection

The authors conducted 19 semistructured interviews, at which point theoretical saturation was reached. Participants were from 19 organizations in 11 industrial sectors, including research/exploration, telecommunications services, industrial services, oil and gas related, equipment and services, software and IT services, commercial printing services, insurance, food and beverage, banking and investment services, and logistics, which were categorized using the Reuters categorization system (Reuters 2013); and the interviews spanned four countries (Switzerland, USA, UK, and Germany). The participant roles included CTO director/program manager, PMO lead, project manager, delivery IT manager, systems engineer lead, head of R&D research, CFO/COO, and general manager.

The demographic information is summarized in Appendix A (page 107). The level of the interviewees varied from project manager, program manager, and PMO lead to CTO and COO; therefore, some relevant information specifically regarding the usage of the methodologies and their purported strengths and weaknesses needed to be considered against the level of the interviewee.

The interviews were semistructured and lasted between 60 and 90 minutes. Interview notes and recordings were written up and compared for cross validation. When additional questions or clarity were required, follow-up was done using Skype sessions and email.

6.3.4 Data Analysis Method

Every interview was recorded, and notes were taken at the same time. Each interview was transcribed within a day, as recommended by Miles and Huberman (1994). The transcripts were read several times in order to become familiar with the data in detail (Eisenhardt 1989) and then coded into the NVIVO software tool according to the topics and questions. Data reduction, data display, and deducing conclusions and verification techniques were used to analyze the interview data following the guidelines outlined by Miles and Huberman (1994).

As the interviews progressed, a methodology categorization was developed to show the positioning of methodology in terms of origins and levels of customization.

6.3.5 Validity and Reliability

Once the findings from the interview notes and transcripts were verified to ensure that the findings were credible, the checklist by Miles and Huberman (1994, pp. 278–279) was used to inspect the analysis of the processes and the results. The reliability and validity were assured by considering the following:

- **Reliability.** Interview protocols were reviewed by peers and the data collected from a spread of industries and countries (USA, Switzerland, Germany, and the UK). Identified patterns were cross-validated for reliability.
- **Internal validity.** Concept validity was provided through the theoretically derived model, which was built on existing literature, and from which the propositions were drawn. Construct validity was achieved through convergence of the interviewee data.

6.4 Analysis and Results

This section is structured into two parts: findings relating to Proposition 1, and findings relating to Proposition 2.

6.4.1 Findings Relating to Proposition 1

- **Defined term for parts of a methodology—elements.** The unit of analysis is the relationship between methodology and project success. Every methodology comprises a number of parts or elements. From the authors' perspective, the term *parts* does not seem to be appropriate; therefore, a commonly understood term was required. The literature does not provide a suitable term. This may be due to a research focus on the impact of a methodology as a whole, on project success, or on the use of one part of the methodology, such as scheduling or risk management, and its impact on project success. This study looks at all of the parts of the methodology in which the independent variable—project methodology—includes processes, tools, techniques, methods, capability profiles, and knowledge areas. Interviewees were asked to provide a term that encompasses all parts of their methodology; the majority believed that *elements* was the appropriate term to use.

- **Project success.** To understand what project success is, success criteria need to be defined; otherwise, success could mean something different to each person. The interviewees were asked whether project success is defined within the organization. None of the 19 interviewees said their organization has a standard definition for project success. When asked how their performance was evaluated on projects, the majority mentioned time, cost, scope, and sometimes customer satisfaction. For the research organizations, success was described in terms of the number of ideas, the number of ideas moved to development, and the number that were industrialized.

Table 6.2 Impact of Project Methodology on Project Success Characteristics

Characteristics of Project Success	Number of Interviewees Mentioned
Cost	17
Time	14
Scope	11
Customer satisfaction	9
Quality of deliverable	3
Ideas developed	1

The interviewees were asked whether project methodology within the organization impacts the characteristics of project success. Table 6.2 shows that project methodology does impact the project success characteristics, where the highest references were to time, cost, and scope.

One of the interviews stated "Yes, 100%" and then described method elements that, if not executed correctly, would impact the characteristics of project success: "Requirements management not followed through results in insufficient scope development and insufficient project governance around changes."

These findings support Proposition 1: There is a positive relationship between project methodology and project success.

6.4.2 Findings Relating to Proposition 2

Evolution of project methodologies. Project methodologies have been evolving and adapting over the years through need and perceived impact on project success. Morris and Pinto describe this by writing, "It's time to move on project management from a rather tired and dated positivist or normative origin stemming with its roots firmed in engineering companies to perhaps where it needs to reflect much more in a complex reality, such as organizational change-type projects where interpretive views of the reason for change are more appropriate" (Morris and Pinto 2004). The international standards, such as PMI's *A Guide to the Project Management Body of Knowledge®* (*PMBOK® Guide*) (PMI 2018) and the UK Office of Government's Commerce (OGC) PRINCE2®, are updated every few years and include extensions for government, construction, defense, and the software industries.

Figure 6.2 was developed from the interviews to help structure the source and levels of methodology customization. Of the organizations interviewed, 65% of the methodologies were based on an international standard, and of these organizations, 75% customized the international standards to varying degrees. It is interesting to note that 35% of the organizations interviewed had more than one methodology that was customizable per project type. Two of the interviewees in the software consulting business explained that their organizations had over 40 methodologies that were used for different applications, industries, and project types. This shows the apparent need and benefit for some organizations to employ specialized methodologies according to application, project type, and business area. None of the organizations interviewed indicated that their project methodologies were customized at the level of the project team and skills; however, this may be implicitly done by the project teams in organizations that allow further levels of customization.

These findings indicate that environmental factors have a moderating effect on the relationship between project methodology and project success, because the interviewees' organizations are invested in creating and maintaining customized/tailored methodologies. One of the interviewees said that, "The company culture impacted whether the elements of the methodology were used or not; typically change management, risk management, and issue management were not used or done properly," reiterating the moderating effect of environmental factors.

Impact of the project environment (including project governance) on the relationship between methodology and the characteristics of project success. Environmental factors are conditions or things that are outside of the immediate control of the project team which influence, constrain, or direct the project, program, or portfolio (PMI 2018). These factors create the context for

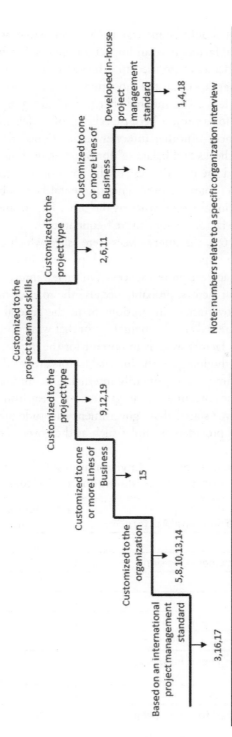

Figure 6.2 Methodology Origins and the Levels of Customization

the project and how it should be managed. The interviewees were asked which internal environmental factors have an impact on the relationship or the way in which methodology elements are used to achieve process success. Referring to Table 6.3, project governance was the most frequently mentioned environmental factor. One interviewee commented that it is challenging to get the right governance structure for a project because "clients often felt they did not have the time for governance." Another interviewee mentioned that project governance provided the "checks and balances" for other environmental factors such as politics, power, and the effectiveness of the sponsor. Governance may have been raised more times than the other environmental factors because it can be considered an institutional factor, whereas the other factors are associated with individuals, and these factors change more frequently.

Referring to Table 6.4, the interviewees were asked which external environmental factors have an impact on the relationship or way in which methodology elements were used to achieve project success. Only a few interviewees mentioned external environmental factors, probably because the roles of these interviewees were in supporting government institutions or in the general consulting area. There was no single external environmental factor that was more prominent than others. These external factors were as important for the impact of the effectiveness of the project methodology as the internal factors are for those interviewees whose project environment was primarily internal. One interviewee stated that "when dealing with the government, things are never clear from the beginning"; and another interviewee stated that "government can suddenly change priorities immediately" and provided examples such as the government shutdown or

Table 6.3 Internal Environmental Factors Impacting Project Management Success

Internal Environmental Factors	Number of Interviewees Mentioned
Governance	10
Political-senior management decisions	4
Leadership maturity	4
Culture	4
Skills and resource constraints	4
Pressure to reduce project costs	4
Sponsor understanding need for a project methodology	3
Understanding requirements	3
Understanding the need for good project management	1

**Table 6.4 External Environmental Factors
Impacting Project Management Success**

External Environmental Factors	Number of Interviewees Mentioned
Regulatory and legal requirements	2
Client culture	2
Governance structure(s)	1
Client's understanding of project management	1
Changes in policy, priorities	1
Funding	1

regulations on hiring. Interviewees working in consulting positions for companies raised the issue of "client culture" that is not conducive for projects, in addition to a lack of understanding of what is required in project management.

Both internal and external environmental factors act as moderating variables.

Methodology elements impacted by the moderating effect of environmental factors. The interviewees were asked which elements of their project methodology(s) that relate to project success were impacted (moderated) by environmental factors, but without specifically focusing on any one environmental factor. Referring to Table 6.5, stakeholder management and change management were the top two methodology elements that were impacted by environmental factors. One of the interviewees mentioned that the culture of their organization was to show good results, and reports were changed to reflect this. Another interviewee working in an external role responded to the question by stating, "Methodology elements are impacted 100%, first by insufficient scope development; and second by insufficient project governance around change management." One interviewee made an interesting point by stating that the "organization's project methodology was specifically developed in-house and takes into account company culture to reflect the context of the organization." The implication is that culture may also have a direct impact on methodology as well as being a moderator on the relationship between project methodology and project success.

The interviewees were then asked to take project governance as the environmental factor and identify which elements of their project methodology(s) related to project success were impacted (moderated) by project governance. Referring to Table 6.6, cost and stakeholder management were mentioned the most as being impacted by project governance. The references to the impact on the stakeholder management methodology element were positive and

Table 6.5 Methodology Elements' Relationship to Project Success Impacted by Environmental Factors

Metholodology Elements Impacted	Number of Interviewees Mentioned
Stakeholder management	11
Change management	10
Risk management	8
Cost estimating	8
Cost control	7
Testing	3
QA—assessments	3
Benefits management	3
Configuration management	2
Issue management	2
Technical estimating	1
Reporting	1
Technical reviews	1
Stagegate reviews	1
Procurement	1

negative. On the positive side, some interviewees, primarily in the consulting area, designed stakeholder involvement and decision making around the project governance structures to ensure a full alignment with all decisions made. On the negative side, others mentioned that stakeholders were excluded from critical parts of the project management life cycle mainly because of ill-fitting project governance structures.

One of the interviewees involved in software consulting said that, "During the process of bidding for the work, which can last for months, the project governance structures are well defined before the contract is signed so that any decision made to deviate from the plan in terms of using the elements of the methodology are agreed upon in writing by the various levels." Another interviewee explained that, "Project governance is used at the setup of the project where the project manager is required to justify why specific elements of a methodology will not be used." This was the only reference provided where every methodology element must be used unless there is justification for not using an element.

Two methodology elements—cost estimating and cost control—were raised in both the positive and negative contexts. Projects in control use proper cost estimation (using the cost control element), and projects out of cost control are so because of lack of project governance impacting the use of the cost control element.

Table 6.6 Methodology Elements' Relationship to Project
Success Impacted by Project Governance

Governance Moderation Effect on the Unit of Analysis	Number of Interviewees Mentioned
Stakeholder management	8
Cost control	7
Change management	5
Cost estimating	5
Risk management	3
Testing	2
QA—assessments	2
Benefits management	2
Issue management	2
Configuration management	1
Reporting	1
Technical reviews	1
Technical estimating	0
Stagegate reviews	0
Procurement	0

The findings from the interviews support Proposition 2: There is a moderating effect of the project environment on the relationship of methodology usage and project success.

6.5 Discussion

The interview results showed the importance of project methodologies and their elements, as they directly impact the characteristics of project success. This is consistent with the finding in the literature at the methodology level (Belassi and Tukel 1996; Pinto and Prescott 1988; Zwikael and Unger-Aviram 2010).

An applied project methodology consists of a number of elements that collectively impact the characteristics of project success. The interviewees mentioned 15 methodology elements that impacted project success, and of these 15 elements, they discussed 13 elements whose effectiveness in supporting project success they believed was influenced by project governance.

Therefore, the nature of the relationship between project methodology elements and project success seems to be contingent on the project environment, notably project governance. Discussions on the impact of project governance were mainly from a positive perspective. Therefore, the influence of project governance on the elements of a project methodology was based on the premise that project governance was supportive of project success.

However, if project governance was misaligned or suboptimal with respect to supporting the project, the positive impact of trying to apply a methodology would be reduced or even detrimental to project success. This can be explained by the effect of Type 1 and Type 2 errors (Atkinson 1999), which was not directly discussed during the interviews but was implied by the discussion of suboptimally implemented methodology elements. For example, one of the interviewees stated that a project with a poor governance structure resulted in suboptimal and delayed decision making (a Type 1 error), which impacted project success.

Not all methodology elements are equal, meaning some of the methodology elements may have a greater impact on project success than others. The literature describes these as *project success factors* (Pinto and Slevin 1987). Success factors may be linked to one or more underlying methodology elements, but the determination of which elements or groups of elements are highly correlated to project success requires further research.

The literature is divided on whether standardized or customized methodologies provide higher project success rates (Lehtonen and Martinsuo 2006) but does not cover the topic of evolving or adapting methodologies owing to new innovations or environmental changes. All methodologies should evolve, including standardized methodologies, to ensure the closest environmental fit to the project environment with the most appropriate methodology elements.

We continue with the theme of evolving and adapting methodologies but take a different perspective on project methodologies to provide additional insight. Some elements of a methodology have a greater impact on the characteristics of project success than others. A natural-science comparative model by Joslin and Müller (2013) compares project methodology elements to the genes of an organism. The genes of an organism are the building blocks of the organism (including the observable characteristics) called a *phenotype* (Malcom and Goodship 2001). Genes are switched on and off throughout the life of an organism, which the authors argue is the same concept as the elements of a methodology being applied (switched on), when required, to a project throughout its life cycle, then switched off when not required.

The nature–social-science comparative reifies a project methodology that is considered as the core makeup of a project and, therefore, is responsible for the switching on and off of methodology elements. The project manager is considered to be an environmental variable. Some of the genes in an organism are highly pleiotropic, meaning their impact can be seen in the organism's phenotype—for example, hair color, eyes, and height (Stearns 2010). The comparative explains that the same is true for elements of the applied project methodology.

The highly pleiotropic methodology elements noticeably impact the characteristics of project success.

Returning to the interviews, some interviewees discussed the impact on project success of certain methodology elements when they were not used because of the impact of environmental factors. The examples given included change requests, risk management, and deliverable sign-off procedures. The resulting consequences on the characteristics of project success included increased costs, quality issues, and reduced customer satisfaction. These examples could be indicators of highly pleiotropic methodology elements. This alternative perspective of a natural-science comparative may provide new insights that would not be possible using a social science perspective.

Humans like to rationalize and standardize what others falsely assume as progress (Habermas and Lawrence 1990), whereas nature is for specificity and optimization to the environment (Dawkins 1974). Methodologies exist within a socially constructed world, but one could argue that these methodologies coexist within a natural science world—a world that contends with survival of the fittest (Dennett 1996); therefore, the concept of methodologies is likely to show the characteristics of both worlds. For example, some methodology elements may be considered common or core versus other elements that are considered contingent or more impacted—that is, influenced by the environment. Certain elements should be clear and common—for example, scope definition, clear project mission/goals, good cost management, time management, etc. Other elements may be more variable—for example, acceptance of variation/changes and stakeholder engagement, etc.

Perhaps referring to core (standardized) and subsidiary (unique) methodology elements, wherein the latter is more influenced by the environment, would be an interesting perspective of a core makeup of a project. Both core and subsidiary elements are under the effect of project governance and, depending on the governance inclination, may converge or diverge with other methodology elements, thereby challenging whether standardized, partly standardized and partly customized, or fully customized methodologies best achieve project success.

Contingency theory within the field of project management offers insight into how best to adapt project management practices within a given environment to meet the project management goals (Hanisch and Wald 2012). Contingency theory applies to selecting and customizing the project methodology according to the environment. The findings from the interviews show that the effectiveness of the methodology to achieve project success is moderated by the project environment. Project governance was the most frequently mentioned environmental factor impacting the effectiveness of the applied project methodology. Examples were given of ill-fitting project governance structures that impacted the ability to follow procedures, to obtain resources, and to finalize requirements, test strategies, and quality assurance. The findings did not go so far as to suggest actions to enhance the positive aspects and minimize the negative aspects of the environmental project governance factor.

This study's findings show that methodologies should be viewed at the methodology-element level, at which the elements collectively impact project success. The nature of the relationship between the methodology elements and project success is dependent on the project environment, which impacts the effectiveness of the elements to such a degree that Type 1 and Type 2 errors start to occur (Atkinson 1999). All the organizations interviewed have either a methodology based on an international standard that has been customized in some way to the organization in varying degrees or a project methodology developed in-house.

Understanding the origins of a project methodology highlights the significance of methodology customization, which may not be apparent when the origins of the methodology are ignored or not understood. Therefore, project methodologies that are termed as standardized may have gone through several iterations of customization, because they were first implemented based on the premise that methodology effectiveness is contingent on the project environment.

6.6 Conclusions

This qualitative study interviewed 19 project, program, and senior IT managers from 11 industries across four countries, all of whom have detailed knowledge of their organization's methodology(s). A deductive approach was used to validate a theoretically derived research model.

The findings show that there is a positive relationship between project methodology elements and the characteristics of project success; however, the influence of the project environment, notably project governance, can influence the effectiveness of this relationship. The findings also show that missing or misaligned governance structures can introduce Type 1 and Type 2 errors.

Contingency theory within the field of project management offers insight into how to best adapt project management practices within a given environment to meet the project management goals. This study has helped to achieve the research aims to qualitatively validate the constructs of the research model, gain agreement on the use of the terms *methodology elements* and *project success*, and gain additional insights, such as the importance of understanding the methodology source and levels of customization.

6.6.1 The Practical Implications

When an organization is considering the replacement of an institutionalized project methodology (including a project methodology with derivatives), it is important to understand the context and how it is reflected in the incumbent methodology. With this information, an informed decision can be made.

For project managers using a project methodology, there is a risk of suboptimal project performance, because the effectiveness of methodology elements may be negatively impacted by environmental factors. The project manager should understand which project methodology elements are the foundation for success factor variables and understand and manage the potential reduced effectiveness of those project methodology elements that could increase the risk of project failure.

6.6.2 The Theoretical Implications

Project governance plays a major role in the moderating effect of project methodology performance, and contingency theory is applicable to methodology selection and its customization according to the project environment.

A project methodology's effectiveness is impacted continuously by the project environment, in which the result can be seen in the characteristics of project success. Viewing a project methodology from a natural science perspective may bring new insights into the behavior and effectiveness of methodologies in different contexts.

6.6.3 Strengths and Limitations

The study collected data from various industries and countries to theoretically derive the research model. The depth of the interview discussions and the experience of the interviewers helped to provide rich data, generating new insights which would not have been possible from an online survey.

This study is based on interviews of a small sample size; therefore, the results cannot be generalized.

6.6.4 Future Research

- To better understand how generic versus customized methodologies are impacted by environmental factors—for example, is there a commonality between the environmental factors that impact the elements of a generic methodology and those environmental factors that impact a highly customized project methodology?
- To understand if project type impacts the relationship between project methodology and project success; and further, to determine whether different environments impact the completeness of an organization's methodology. In other words, are some organizations' methodologies more comprehensive than others, and, if so, what are the implications?

6.6.5 Contributions to Knowledge

The value of this study lies in the following:

- A project methodology should be seen as a collection of methodology elements, all of which impact the characteristics of project success and wherein some methodology elements are the foundation of success factors.
- Identification of environmental factors, especially project governance, which impact the relationship of project methodology and the characteristics of project success.
- To provide empirical data for a prestudy in a new field of study using a new method. A natural- to-social-science comparative was created, comparing project methodology elements to genes of an organism (Joslin and Müller 2015a). The results of this study, in conjunction with a greater study, will be used to determine the validity of the new comparative.

Appendix 6A: Interview Protocol

1. Nature of the organization and types of projects within your organization

- What types of business activities are carried out in your organization?
- What types of projects are carried out in your organization?
- What categories of projects are undertaken? Compare with Table A.1.
- What is the criterion to judge project size in terms of small, medium, and large in your organization?

2. The project methodology(s); how it was originally developed and evolved, what are the project types supported, what are its strengths and weaknesses?

Please describe the project methodology or methodologies your organization uses, including whether it is based on an international standard such as PRINCE2®, Prompt, or the *PMBOK® Guide*?

- If the methodology was based on an international standard, then was the methodology tailored/customized to your business, and, if so, was it tailored/customized per project type or per business section?
- If the methodology was developed within your organization, was it developed for a specific product or service? Please describe its background.
- Are there derivatives of the methodology for different types of projects or business areas and, if so, describe why?

Please describe the strengths and weaknesses of your project methodology:

- Are there certain types of projects that your methodology is less or more suited to?
- Does your project methodology evolve to meet organizational needs, and, if so, how does it evolve? Also, who is responsible for its evolution?
- What would you recommend to improve the value of your organization's methodology?
- Looking at the methodology, what word would you use to describe the parts of the methodology (hierarchical breakdown) in a generic sense?
- Does your project methodology for any given project type integrate the "how to build" something with the "what to build," or is the "what to build" (requirements specs) kept separately?
- Would there be any advantages or disadvantages in combining a methodology, and what needs to be built into one integrated approach?

Table 6A.1 Categories of Project Types

Application	Project Type	Internal / External	Urgency[a]	Innovation[a]	Size	Technology[a]	Complexity[b]
Engineering	Research	Internal	Low	None	Small	Low	Task
IT	Consulting	External	Medium	Incremental	Medium	Medium	Simple project
Organizational change	Development	Both	Time-critical	Breakthrough	Large	High	Normal project
Marketing	Operations		Crisis	Transformational		Extreme	Daunting project
Finance	Decommissioning						Program
	Organizational change						
	Service improvement						
	Service development						
	Service decommissioning						

[a] The ranges of urgency, innovation, and technology are taken from Shenhar's Diamond model (Shenhar and Dvir 2007).

[b] The ranges for complexity are taken from PRINCE2® (TSO 2009).

3. Project success (success)

- Is there a definition of project success in your organization?
- Is there a definition of project success for your project(s)?
- Are there any numbers published on project success rates?

4. Impact of a project methodology on project success

- Have you observed the project methodology, including how its elements impact the characteristics of project success?

5. Project governance paradigm based on Müller (2009) and how it relates to the goals of the organization/shareholders

Background. The corporate governance of an organization can be modeled on a continuum from shareholder orientation to stakeholder orientation. In shareholder-oriented companies, all decisions are driven by the underlying desire to maximize the wealth of the organization's shareholders. In stakeholder-oriented companies, there is still a need to create profit to satisfy the needs of the shareholders, but this is only one of a variety of stakeholder groups.

- Where on this continuum would you place your organization?
- Is there a management philosophy with emphasis on always getting personnel to follow the formally laid-down procedures? Or is there a strong emphasis on getting things done, even when this means disregarding formal procedures?
- Is the project manager responsible for time, cost, budget, and/or any other measure?
- Is the reason that the project manager is responsible or not for something due, in some way, to the governance paradigm used within your organization?

6. Impact of the project environment (including governance) on the relationship between methodology and the characteristics of the project success

- Which environmental factors have an impact on the relationship among methodology elements or the way they are used to achieve process success?
- Consider the governance paradigm impacting your project(s). How has governance, as an environmental factor, impacted the relationship or manner in which methodology elements are used to achieve process success?

7. Anything else you think is important to add?

Chapter 7

Relationships Between a Project Management Methodology and Project Success in Different Project Governance Contexts

Coauthored with Ralf Müller
BI Norwegian Business School, Norway

7.1 Introduction

This study looks at the relationship between the use of a project management methodology (PMM) and project success, and the impact of project governance context on this relationship. A cross-sectional, world-wide, online survey yielded 254 responses. Analysis was done through factor analysis and moderated hierarchical regression analysis. The results of the study show that the application of a PMM accounts for 22.3% of the variation in project success, and PMMs that are considered sufficiently comprehensive lead to higher levels of project success than PMMs that need to be supplemented for use by the project manager.

Project governance acts as a quasi-moderator in this relationship. The findings would benefit project management practitioners by providing insights into the selection of PMM in different governance contexts. Researchers would benefit from insights into PMM's role as a success factor in projects.

Project success is one of the most researched topics in project management, but the meaning of the term *success* varies substantially (Judgev and Müller 2005). Cooke-Davies (2002) makes the distinction between *project success,* which is measured against the overall objectives of the project and accomplished through the use of the project's output, and *project management success,* which is measured at the end of the project against *success criteria,* such as those relating to internal efficiency—typically cost, time, and quality (Atkinson 1999). The accomplishment of these criteria can be influenced throughout the project life cycle through *success factors* (Müller and Turner 2007b).

One of these factors is the project management methodology (PMM), which is meant to enhance project effectiveness and increase chances of success (Vaskimo 2011). Thus, PMMs were developed to support project managers in achieving more predictable project success rates. However, the extent to which this objective is reached is unknown, because projects still fail to reach their goals (Lehtonen and Martinsuo 2006; Wells 2013), and a quantification of the impact of PMMs on project success is still missing. Examples of internationally recognized PMMs include PRINCE2® from the Office of Government Commerce (OGC 2002), The System Development Life Cycle (SDLC) (Ruparelia 2010), and Erickson's PROPS (Ericsson 2013), whereas PMI's *A Guide to the Project Management Body of Knowledge® (PMBoK® Guide)* is a body of knowledge and not a methodology (PMI 2018).

The project management literature distinguishes between standardized versus customized PMMs (Crawford and Pollack 2007; Curlee 2008; Fitzgerald, Russo, and Stolterman 2002; Milosevic and Patanakul 2005; Shenhar and Dvir 2002) and is divided on whether standardized PMMs, customized PMMs, or a combination of both enhances project effectiveness, hence leading to a higher chance of project success (Curlee 2008; Milosevic and Patanakul 2005; Shenhar and Dvir 1996).

A related perspective is the comprehensiveness of a PMM and its impact on project success (Fortune et al. 2011; Wells 2013; White and Fortune 2002). The premise of being able to standardize and/or customize a methodology is the underlying assumption that the PMM will then become comprehensive—that is, sufficient for any given project.

When an organization's PMM is incomplete or limited (missing methodology elements), project efficiency, quality, and ultimately the probability of project success will be impacted. Fortune and White (2011) showed that more than 50% of the respondents in their study experienced limitations using PMMs. Among the most often mentioned were limitations in methods, processes, tools, and techniques. A method is a set of procedures, to be used by humans, for selecting and applying a number of techniques and tools in order, efficiently, to achieve the construction of efficient artifacts (Bjorner and Druffel 1990).

Simply put, a method is what is applied in a particular situation, and a methodology is the sum of all methods and the related understanding of them.

Wells (2013) and Joslin and Müller (2016b) found that PMMs vary in completeness and appropriateness from organization to organization. Some are considered inadequate for certain types of projects. These reported issues suggest that it is not sufficient to look at a PMM as a whole, especially as every PMM is a heterogeneous collection of practices that vary from organization to organization (Harrington et al. 2012). In this chapter, the elements of a PMM are first defined, and then they are investigated as to their collective impact on project success in governance contexts.

Governance pervades organizations. "Corporate governance encompasses all work done in an organization, and thus governs the work in traditional line organizations, plus the work done in temporary organizations, such as projects," and project governance is a subset of corporate governance (Müller et al. 2013, p. 26). The definition of corporate governance, which has been taken from the Organization for Economic Co-operation and Development (OECD), is as follows:

"Involving a set of relationships between a company's management, its board, its shareholders and other stakeholders [. . .] and should provide proper incentives for the board and management to pursue objectives that are in the interests of the company and its shareholders and should facilitate effective monitoring" (OECD 2004, p. 11). Corporate governance influences project governance as an oversight function which collectively encompasses the project lifecycle to ensure a consistent approach to controlling the project with the aim of ensuring its success.

Since 2005, the literature on governance in the realm of projects has grown exponentially (Biesenthal and Wilden 2014). However, the role of PMMs in different governance contexts has attracted very little attention in the past. A notable exception is the study by Joslin and Müller (2016b), which showed that project governance—which is defined as "the use of systems, structures of authority, and processes to allocate resources and coordinate or control activity in a project" (Pinto 2014, p. 383)—may influence the effectiveness of using PMMs to achieve project success. A further refinement of this result is indicated through (1) a quantitative approach that allows for generalizable results, and (2) more granularity in the identification of the particular elements of a PMM that relate to project success.

The aim of this study is to further investigate the relationship between a PMM and its elements with project success, and how this relationship is impacted by different project governance contexts. Consequently, the following research question is proposed:

What is the nature of the relationship between a PMM and project success, and is this relationship influenced by project governance?

The unit of analysis is the relationship between the PMM and project success. In line with the nature of the research question, the study takes a contingency theory perspective.

The results of the study will provide a better understanding of an organization's PMM in terms of the impact of a PMM on project success and how different project governance contexts influence the selection, effectiveness, and comprehensiveness in the use of PMMs.

These findings help organizations to understand how to align their PMMs to optimize effectiveness in use, which should result in higher project success rates and reduce the complaints about ill-fitting PMMs.

This chapter continues by reviewing the related literature, which is followed by the methodology and analysis sections. The chapter finishes with a discussion and conclusions.

7.2 Literature Review and Hypotheses

This section reviews the literature on project success, project PMMs, and governance from which the hypotheses are derived and describes contingency theory as the theoretical perspective.

7.2.1 Project Success

Since the 1970s, academics have tried to understand what project success is and which factors contribute to it (Ika 2009). However, its meaning is still not generally agreed upon (Judgev and Müller 2005). Project success is a multidimensional construct that includes both the short-term project management success *efficiency* and the longer-term achievement of desired results from the project—that is, *effectiveness and impact* (Judgev, Thomas, and Delisle 2001; Shenhar, Levy, and Dvir 1997).

To achieve a common understanding of what project success is, it should be measurable and therefore defined in terms of success criteria (Joslin and Müller 2016b). The understanding of project success criteria has evolved from the simplistic triple constraint concept, known as the iron triangle (time, scope, and cost), to something that encompasses many more success criteria (Atkinson 1999; Judgev and Müller 2005; Müller and Jugdev 2012; Shenhar and Dvir 2007). Measurement models for success that are applicable for different types of projects or different aspects of project success were developed by Pinto and Slevin (1988a), Shenhar et al. (2002), Hoegl and Gemuenden (2001), and Turner and Müller (2006).

At the same time, project success factors has become a popular theme in research (e.g., Belassi and Tukel 1996; Cooke-Davies and Arzymanow 2002; Pinto and Slevin 1988a; Tishler et al. 1996; White and Fortune 2002). Factors can be categorized into environmentally related (meaning, where the project resides) (Fortune and White 2006; Hyväri 2006; Jha and Iyer 2006), people-related (Tishler et al. 1996), processes- and tools-related (Jessen and Andersen 2000; Khang 2008; Shenhar et al. 2002), and just context-related, meaning two or more categorizations (Sauser, Reilly, and Shenhar 2009). In the absence of a formal definition for project context, the definition of the term *context* has been adapted from Abowd, Dey, and Brown (1999): "Project context is any information that can be used to characterize the situation of [a] project which includes physical and mental aspects. The physical aspects of project context include previous projects as well as the project environment where the project actually resides, whereas the mental aspects includes social, emotional, or informational states."

Schultz, Slevin, and Pinto (1987) suggested that the relative importance of success factors varies over the project life cycle. Shenhar et al. (2001) described the importance of success factors not just on the project life cycle but also on the product life cycle from project completion to production, and then to preparation for project/service replacement. Researchers soon realized that success factors without structure, grouping, and context would result in increased project risks; therefore, success factor frameworks were introduced (Judgev and Müller 2005). Pinto developed a success framework covering organizational effectiveness, technical validity, and organizational validity (Pinto and Slevin 1988b). Freeman and Beale's (1992) success framework included efficiency of execution, technical performance, managerial and organizational implications, manufacturability, personal growth, and business performance. Shenhar et al. (2001) described that no one-size-fits-all exists by using a four-dimensional framework, showing how different types of projects require different success factors, determined by the strategic nature and the short- and long-term project objectives.

Khan, Turner, and Maqsood (2013) developed a model of success factors derived from a literature review of the past 40 years. Their model offers a balance between hard and soft factors and measures success using 25 variables organized in five dimensions. The model contains the three criteria for the iron triangle (Dimension 1) plus four additional project success criteria dimensions:

1. Project efficiency
2. Organizational benefits
3. Project impact
4. Stakeholder satisfaction
5. Future potential

Table 7.1 (starting on page 118) contains the list of success criteria variables (questions). Their model was selected for this study because it is based on the latest literature, which is a superset of the success criteria from the leading researchers on project success.

Project success is the dependent variable in the research model.

7.3 Project Management Methodologies (PMMs)

Forty years ago, the first formal PMMs were set up by government agencies to control budget, plans, and quality (Packendorff 1995). Two of the main topics of PMM research involve the context of standardized versus customized PMMs and the comprehensiveness of a PMM.

The literature is split on whether standardization, which implies little environmental context; customization, which implies context; or a combination of both, which implies some context, lead to a higher chance of project success.

- **Standardization.** A PMM and its processes have been referred to as *organizational processes*, implying that they have degrees of standardization (Curlee 2008). "Owners" of project management practices often perceive projects as a means to attain corporate goals and, therefore, follow the path of corporate control and standardization (Packendorff 1995). Project management offices (PMOs) are focused on standardizing organizational PMM and project management per se (Hobbs, Aubry, and Thuillier 2008).
- **Customization.** Shenhar and Dvir (1996) were the first proponents of customization in showing that projects exhibit considerable variation, which, at that time, went against the literature trend, which assumed that all projects were fundamentally similar. In repeating Shenhar et al.'s mantra, Wysocki (2011) stated that the often-used term "one size fits all" does not work in project management. This is supported by Payne and Turner (1999), who found that project managers often report better results when they can tailor procedures to the type and size of the project they are working on or the type of resource used on the project. Russo and Stolterman (2002) noted that the most successful PMMs are those developed for the industry/organization aligned to the context factors.
- **Combination of standardization and customization.** A contingency approach was suggested by Milosevic and Patanakul (2005), in which it made sense to standardize only parts of the PMM in an organization. Aubry et al. (2010) found that the more experienced PMOs were using methods derived from agile PMMs that allowed flexibility in the processes and PMM. Turner, Ledwith, and Kelly (2010) noted that organizations vary in size, as do their PMM requirements.

The literature on PMMs is divided on whether standardized or highly customized PMMs are more effective in supporting project success, but the research implies the importance of context, albeit in varying degrees. In this chapter, we look at the impact of context on the effectiveness of a PMM.

Independent of whether a PMM is standardized, customized, or a combination of both, when the organization's PMM is incomplete or is limited, the efficiency of the project will be impacted. Joslin and Müller (2016b) and Wells (2012) found that PMMs vary in completeness and appropriateness from organization to organization, in that some are considered inadequate for certain types of projects. White and Fortune (2002), using a survey on project management practices, reported that very few methods, tools, and techniques were used; and for the ones that were used, almost 50% of the respondents reported drawbacks to the way these were deployed. Fortune and White (2011) stated that 27% of respondents experienced limitations with in-house PMMs, and 57% of respondents experienced limitations with other PMMs.

These reported issues suggest that it is not sufficient to look at the PMM as a whole, because every PMM is really a heterogeneous collection of practices that vary from organization to organization (Harrington et al. 2012). A common understanding is required to understand what the elements (or parts) of a PMM are, and what their impact is on project success. With this information, the issues reported on PMM limitations can be further investigated. We look at defining the elements of a PMM and determine their impact on project success in different contexts.

To understand what constitutes a PMM, several international standards were reviewed. The Project Management Institute (PMI 2018) describes a PMM as "a system of practices, techniques, procedures, and rules," whereas PRINCE2 from the UK is not described as a PMM, but rather as a method that contains processes but not techniques.[1] Ericsson's PROPS PMM from Sweden does not call itself a PMM but a model, wherein the model describes all of the project management activities and documentation (Ericsson 2013). In the absence of a consistent description for the elements of a PMM, this study uses the definition of PMM elements from Joslin and Müller (2014a), which defines PMM elements as processes, tools, techniques, knowledge areas, and comprehensive capability profiles.

A PMM should take into account different levels of scope and comprehensiveness, in which the term *comprehensiveness* is taken to mean *including or dealing with all or nearly all elements or aspects of something* (OxfordDictionaries

---

[1] The Office of Government Commerce (OCG) leaves it up to the project manager to decide on the relevant techniques to use during the project life cycle.

Table 7.1 Project Success Questions

My last project was successful in terms of:	Project Success Achieved				
	Not Successful	Slightly Successful	Moderately Successful	Highly Successful	Very Highly Successful
Completed according to the specification	○	○	○	○	○
Supplier satisfied	○	○	○	○	○
Enabling of other project work in future	○	○	○	○	○
Project achieved a high national profile	○	○	○	○	○
Yielded business and other benefits	○	○	○	○	○
Met client's requirement	○	○	○	○	○
Minimum disruption to organization	○	○	○	○	○
Cost effectiveness of work	○	○	○	○	○
Met planned quality standard	○	○	○	○	○
Adhered to defined procedures	○	○	○	○	○
Learned from project	○	○	○	○	○
Smooth handover of project outputs	○	○	○	○	○
Resources mobilized and used as planned	○	○	○	○	○
Improvement in organizational capability	○	○	○	○	○
Met safety standards	○	○	○	○	○
Minimum number of agreed scope changes	○	○	○	○	○
Motivated for future projects	○	○	○	○	○

Table 7.1 Project Success Questions (cont.)

My last project was successful in terms of:	Project Success Achieved					
	Not Successful	Slightly Successful	Moderately Successful	Highly Successful	Very Highly Successful
Project's impacts on beneficiaries are visible	O	O	O	O	O
Project achieved its purpose	O	O	O	O	O
Project has good reputation	O	O	O	O	O
Finished on time	O	O	O	O	O
New understanding/knowledge gained	O	O	O	O	O
Steering group satisfaction	O	O	O	O	O
Complied with environmental regulations	O	O	O	O	O
End-user satisfaction	O	O	O	O	O
Project team satisfaction	O	O	O	O	O
Activities carried out as scheduled	O	O	O	O	O
Finished within budget	O	O	O	O	O
Sponsor satisfaction	O	O	O	O	O
End product used as planned	O	O	O	O	O
Personal financial rewards	O	O	O	O	O
Met organizational objectives	O	O	O	O	O
The project satisfies the needs of users	O	O	O	O	O
Personal nonfinancial rewards	O	O	O	O	O

2014). PMMs that are not comprehensive are considered incomplete in this study and, therefore, will need to be supplemented during project execution.

Each organization must decide on the level of PMM comprehensiveness, for which the more comprehensive the PMM, the less the need for it to be supplemented when it is applied to a project. In this study, the term *organization's comprehensive PMM* means the implemented PMM within an organization and its ability to support all of the project types without the need to be supplemented with missing elements (Mengel, Cowan-Sahadath, and Follert 2009, p. 33). Some organizations may choose not to invest in a comprehensive PMM or training and instead assume that their project PMM will always need to be supplemented, thereby leaving this decision to the user of the PMM. This is called *supplementing missing elements.*

Irrespective of whether a PMM is supplemented or not, the user may still decide to apply only a subset of the PMM. This is done in an attempt to apply only those elements of a PMM required for achieving the desired project outcome. We refer to this as *applying relevant PMM elements* throughout the chapter.

Studies showed that organizations experience limitations in their PMMs, irrespective of whether it is an in-house or an off-the-shelf PMM (Fortune et al. 2011; White and Fortune 2002). Wells (2013) found that when the selection of PMMs at the organizational level did not address the needs of the departments and projects, project managers would tailor their organizational PMMs specifically for their projects.

The literature review suggests the existence of a knowledge gap regarding the collective impact of a project's PMM elements on project success.

Hypothesis 1: There is a positive relationship between a PMM and project success.

- **H1.1.** There is a positive relationship between a comprehensive set of PMM elements and project success.
- **H1.2.** There is a positive relationship between supplementing missing PMM elements and project success.
- **H1.3.** There is a positive relationship between applying relevant PMM elements and project success.

7.3.1 Project Governance as a Context Factor

Governance influences people indirectly through the governed supervisor and directly through subtle forces in the organization (and society) in which they live and work (Foucault 1980). Governance exists in every facet of life and interacts with laws and contextual frameworks, but it does not determine the actions of the members of a group or team (Clegg 1994). There are various definitions of governance which vary in scope and focus—for example, governance of society, public governance, corporate governance, governance of projects, and project governance. Klakegg, Williams, and Magnussen (2009) define governance as "the use of institutions, structures of authority, and even collaboration to allocate resources and coordinate or control activity in society or the economy."

In projects, governance takes place at different levels—for example, groups of projects, such as programs or portfolios of projects, in which the emphasis is on collective governance, which is viewed as governance of projects (Müller and Lecoeuvre 2014). This differs from governance of individual projects, which we defined earlier in this chapter using Pinto's (2014) definition.[2]

The governance of projects combined with project governance coexist within the corporate governance framework, and both cover portfolio, program, and project management governance (Müller et al. 2014). The literature on project governance addresses several contexts, such as project governance for risk allocation (Abednego and Ogunlana 2006), a framework for analyzing the development and delivery of large capital projects (Miller and Hobbs 2005), NASA-specific framework for projects (Shenhar et al. 2005), governing the project process (Winch 2001), mechanisms of governance in project organizations (Turner and Keegan 2001), normalization of deviance (Pinto 2014), and governance in project-based organizations (functional, matrix, or projectized) (Müller et al. 2014). The literature on governance does not cover either the direct influence of governance on a project PMM or the impact of governance on the nature of the relationship between a project PMM and project success. Hence, there is a knowledge gap in the literature for understanding the impact of project governance on the nature of the relationship between a project PMM and project success.

The reason for considering project governance as the context factor is that corporate governance exists from the point of creation of an organization. Project governance has influenced the way individuals have viewed project management because it provides the structure through which projects are set up, run, and reported (Turner 2006). Therefore, project governance is likely to influence the choices taken in selecting, applying, and evolving a PMM. Project governance may also influence the relationship between PMM and project success, which is one of the hypotheses in this chapter. For these reasons, project governance was selected as the moderator factor for the research model (see Figure 7.1).

---

[2] "The use of systems, structures of authority, and processes to allocate resources and coordinate or control activity in a project" (Pinto 2014, p. 383).

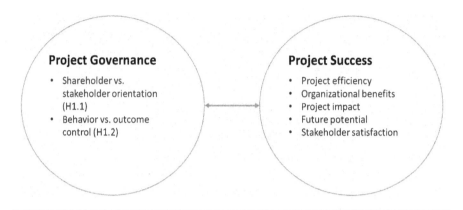

Figure 7.1   Project Governance–Project Success Research Model

To understand the impact of project governance on the relationship between PMM and project success, a framework to categorize each organization's governance is required. Governance models are developed from different perspectives using either a top-down or a bottom-up approach (Klakegg et al. 2009). Top-down approaches are developed from a shareholder–outcome perspective, whereas bottom-up approaches take a process control perspective and can be considered as an extension of a PMM (Müller 2009). This study requires a governance model that considers perspectives of shareholder versus stakeholder, and a "follow the process" behavior approach versus a "get it done" outcome approach. This is required because the governance model perspectives map to the overall objective of a project—that is, a successful outcome, with the objective of a PMM (structured approach to deliver a project), all within an environment that is influenced by shareholders and stakeholders.

Governance models that incorporate topics such as ethics, corporate citizenship, roles, and responsibilities (Dinsmore and Rocha 2012; Renz 2008; Turner 2008; Walker, Segon, and Rowlingson 2008) were excluded because the emphasis of this study is on shareholder–stakeholder and behavior–outcome aspects of the organization. Therefore, the most relevant model was Müller's governance model (2009), which draws on the theories of transaction cost economics, agency theory, and institutional theory using legitimacy to emphasize conformance.

The governance model by Müller (2009) uses categories, called *governance paradigms,* in which an organization governing a particular project fits into one of four paradigms. It addresses corporate governance orientation (shareholder–stakeholder orientation) and the organizational approach to control (behavior versus outcome control). The corporate governance dimension builds on models from Clarke (2004) and Hernandez (2012), who claim that a corporation's governance orientation can be found on a continuum from shareholder to stakeholder orientation. The second dimension "control" represents the control exercised by the governing institution over the project and its manager. This distinguishes between organizational control, which focuses on goal accomplishment by controlling outcomes (e.g., reaching a set of objectives) versus compliance with a focus on employees' behavior (e.g., following a process, such as a project management PMM) (Brown and Eisenhardt 1997; Ouchi and Price 1978; Ouchi 1980).

To address the second part of the research question, based on the literature review we hypothesize that:

Hypotheses 2: The relationship between the project PMM and project success is moderated by project governance.

- **H2.1.** The impact of a comprehensive set of PMM elements on project success is moderated by project governance.
- **H2.2.** The impact of supplementing missing PMM elements on project success is moderated by project governance.
- **H2.3.** The impact of application of relevant PMM elements on project success is moderated by project governance.

7.3.2 Contingency Theory as a Theoretical Perspective

Contingency theory stresses the importance of idiosyncratic structures for organizations, depending on their context (Burns and Stalker 1961; Woodward, Dawson, and Wedderburn 1965). We follow Donaldson's (2001) model of contingency theory in organizations, which explains the effect of one variable (an independent variable) on another variable (a dependent variable) as dependent on a third, a context variable.

A recent bibliographical review of contingency theory in the field of project management showed that it is increasingly used in research, with a noticeable increase since 2005 (Hanisch and Wald 2012). Fitzgerald, Russo, and Stolterman (2002) noted that the most successful PMMs are those developed for industries or organizations that are aligned to context factors. Lehtonen and Martinsuo's (2006) study of project failure and the role of project management PMM concluded that "some contingency variables may have an impact on the relation between PMM and success." This supports the notion of contingency theory, in which the independent variable "PMM" and the dependent variable "success" are influenced by a third variable.

Contingency theory is being used as the theoretical lens for this study to help understand the impact of project PMM on project success in the context of governance paradigms.

7.4 Research Methodology

We took a post-positivist perspective in the sense of Tashakkori and Teddlie (2009), who see post-positivism as "currently the predominant philosophy for quantitative research in the human sciences" (p. 69). Post-positivism "assumes that the world is mainly driven by generalizable (natural) laws, but their application and results are often situational dependent. Post-positivist researchers therefore identify trends, that is, theories which hold in certain situations, but cannot be generalized" (Biedenbach and Müller 2011). Tashakkori and Teddlie (2009, p. 87) suggest that "post-positivists prefer using either quantitatively oriented experimental or survey research to assess relationships among variables and to explain those relationships statistically." This study uses a deductive approach and cross-sectional questionnaire to validate the model shown in Figure 7.1.

7.4.1 Questionnaire Development

Five sets of questions were included in the questionnaire. The first set included information about the last project; the next three sets covered project PMM, governance paradigms, and project success; and the last set collected the respondents' demographic information. The questionnaire followed the suggestions of Cooper and Schindler (2011) to ensure the scales, criteria, and wording were consistent and clear. The questions relating to PMM were developed based on prior work by Joslin and Müller (2014a). The PMM dimensions and questions are shown in Table 7.2. The project context questions were based on the governance paradigms from Müller (2009), which were then operationalized in Müller and Lecoeuvre (2014).

The governance paradigms were selected because they have been used successfully in several project management–related studies and reflect an organization's governance positioning with regard to two continuums: (1) shareholder–stakeholder and (2) behavior–outcome. The project success dimensions were based on Khan and Turner (2013). The five dimensions (project efficiency, organizational benefits, project impact, stakeholder satisfaction, and future potential) cover short- and long-term implications of project success. A pilot test was done with 10 respondents. Based on the feedback, minor wording changes were made for understandability. The pilot answers were not used in the analysis.

Table 7.2 The PMM Dimensions

Comprehensive PMM	The organization's project PMM had a comprehensive set of tools.
	The organization's project PMM had a comprehensive set of techniques.
	The organization's project PMM had a comprehensive set of capability profiles.
	The organization's project PMM had a comprehensive set of knowledge areas.
	The organization's project PMM had a comprehensive set of processes.
Supplemented PMM	I supplemented the organization's project PMM when necessary, with missing tool(s).
	I supplemented the organization's project PMM when necessary, with missing technique(s).
	I supplemented the organization's project PMM when necessary, with capability profiles(s).
	I supplemented the organization's project PMM when necessary, with missing knowledge areas(s).
	I supplemented the organization's project PMM when necessary, with missing process(es).
Applied relevant PMM elements	I applied the relevant tools during the project life cycle.
	I applied the relevant techniques during the project life cycle.
	I applied the relevant capability profiles during the project life cycle.
	I applied the relevant knowledge areas during the project life cycle.
	I applied the relevant processes during the project life cycle.
Achieved expected results	I achieved the project results expected by applying relevant tools.
	I achieved the project results expected by applying relevant techniques.
	I achieved the project results expected by applying relevant capability profiles.
	I achieved the project results expected by applying relevant knowledge areas.
	I achieved the project results expected by applying relevant processes.

The recommendations from Podsakoff and Organ (1986) were followed to minimize potential common methods bias, including confirmed anonymity in the introductory text, different layout and scales, and randomizing of the questions. To avoid biases introduced by the respondents' choice of project—for example, providing information about their most successful project—the survey asked respondents to report on their most recently completed project.

7.4.2 Data Collection

Data collection was performed through a worldwide, cross-sectional questionnaire to collect quantitative data for generalizable results. The respondents were contacted using email with a link to the web survey. In addition, the survey details were placed on project management LinkedIn® forums. An email with the survey link was sent to PMI chapters. Data were collected over a period of 14 days in April, 2014. The following filter question was asked to identify qualified respondents: "Do you have an understanding of your organization's or client's project PMM, where you have been involved as a project stakeholder—that is, someone working in or impacted by projects?" By asking this question, 132 responses were disqualified. This resulted in 254 full responses that could be used for analysis. Responses came from 41 countries, with 24% from Europe, 38% from North America, 22% from Australasia, and 16% from other countries. ANOVA analyses on differences between the early and late respondents, as well as between demographic regions, showed no significant differences ($p = 0.149$ and 0.249, respectively). Average work experience was 22 years, and average project-related work experience was 15 years. Sample demographics are shown in Table 7.3.

The respondents' last project information is shown in Table 7.4. Approximately 48% of the projects were less than 1 million Euros, and 96% of the projects were either of medium to high urgency; 42% of projects were executed in matrix organizations, and only 21% were executed in functional organizations.

7.4.3 Data Analysis Methods

Analysis was carried out following the guidelines from Hair et al. (2010). Data were checked for normality (skewness and kurtosis) within the limits of ± 2 and ± 3, respectively. Eight outliers were removed because one-sample tests showed these cases were significantly different from the other cases.

Exploratory factor analysis using principle component analysis was used on PMM, governance, and success variables to identify underlying structures and reduce the number of variables to a manageable size while retaining as much of the original information as possible (Field 2009). Validity was tested through unrotated factor analysis for each dimension, which also served as the Haman test to exclude common method bias-related issues, as suggested by Podsakoff

Table 7.3 Demographics

Characteristic	N	%	Characteristic	N	%
Sector			**Gender**		
Research & development	31	12.2	Male	194	76.4
Engineering/ construction	46	18.0	Female	56	22.0
Information technology/telecom	120	47.1	Other	1	0.4
Media/arts	9	3.5	Total	251	98.8
Relief aid	16	6.3	Missing	3	1.2
Other	29	11.4			
Total	251	98.4	**Geography—working**		
Missing	4	1.6	North America	96	37.8
			Europe	61	24.0
Position held			Australasia	56	22.0
CIO	3	1.2	Other	38	15.0
CTO	2	0.8	Total	251	98.8
Project portfolio manager	17	6.7	Missing	3	1.2
PMO	10	3.9			
Program manager	65	25.6	**Project-related experience**		
Project manager	82	32.3	1 to 5 years	36	14.6
Team member	24	9.4	6 to 10 years	63	25.6
Architect/advisor	6	2.4	11 to 15 years	53	21.5
QA/audit function	3	1.2	16 to 20 years	45	18.3
Technical stakeholder	2	0.8	20 years plus	46	18.7
Business stakeholder	4	1.6	Total	243	98.8
Other	35	13.8	Missing	3	1.2
Total	253	99.6			
Missing	1	0.4	**Work experience**		
			1 to 5 years	36	14.6
			6 to 10 years	60	24.4
			11 to 15 years	46	18.7
			16 to 20 years	49	19.9
			20 years plus	52	21.1
			Total	243	98.8
			Missing	3	1.2

Table 7.4 Last Project Information

Characteristic	N	%	Characteristic	N	%
Duration of last project			**Urgency of Last Project**		
Under six months	44	17.3	Low	11	4.3
6 months to less than 1 year	67	26.4	Medium	107	42.1
1 to 2 years	76	29.9	High	135	53.1
Over 2 years	66	26.0	Total	253	99.6
Total	253	99.6	Missing	1	0.4
Missing	1	0.4			
			Last Project Executed in the following Organizational Structure		
			Projectized Organization	81	31.9
Level of Last Project Complexity			Functional Organization (Department)	55	21.7
Low	24	9.4	Matrix Organization	106	41.7
Medium	117	46.1	Other	11	4.3
High	111	43.7	Total	253	99.6
Total	252	99.2	Missing	1	0.4
Missing	2	0.8			
Value of Last Project					
Under 500,000 (Euro)	85	33.5			
500,000 to 999,999	37	14.6			
1,000,000 to 4,999,999	61	24.0			
5,000,000 to 50,000,000	43	16.9			
Over 50,000,000	27	10.6			
Total	253	99.6			
Missing	1	0.4			

and Organ (1986). The results for each of the three concepts gave a Kaiser-Meyer-Olkin (KMO) sampling adequacy value of 0.8 or higher ($p < 0.001$), indicating the data's appropriateness for this analysis.

Following Sharma, Durand, and Gur-Arie (1981), hierarchical regression analysis was used to test the relationship between PMM and success (Hypothesis 1) and to test the moderating influence of governance on the relationship between

PMM and success (Hypothesis 2). Finally, a number of ANOVA tests compared the means of three or more groups to determine additional information pertaining to two or more of the research model variables. The results are shown in the following sections.

"Years of project experience" was used as a control variable to filter out spurious effects and improve internal validity by reducing the confounding effect of variations in a third variable that could also affect the value of the dependent variable.

7.4.4 Validity and Reliability

Construct validity was ensured through the use of published measurement dimensions (Joslin and Müller 2014a; Khan et al. 2013; Müller and Lecoeuvre 2014), pilot testing of the questionnaire, and unrotated factor analyses. Content and face validity was achieved by using literature-based measurement dimensions and testing them during the pilot.

Item-to-item and item-to-total correlations below 0.3 and 0.5, respectively, showed internal consistency. Reliability was tested using Cronbach's alpha. All constructs showed reliability, with their respective values over 0.70 (Hair et al. 2010).

7.4.5 Preparation for Operationalization of Variables

- **Project success.** Factor analysis produced a single dimension and reliable factor for project success (KMO 0.930, $p < 0.001$) and a Cronbach's alpha of 0.923.
- **Methodology (PMM).** Operationalization was carried out by using a five-point Likert scale ranging from strongly disagree to strongly agree. The three factors—comprehensive set of methodology elements, labeled MF01-COMPREHENSIVE; supplemented missing methodology elements, labeled MF03-SUPPLEMENTED; and applied relevant methodology elements, labeled MF03-APPLIED—were reliable at 0.75 to 0.77 (Hair et al. 2010) (see Table 7.5).

  Factor analysis with Varimax rotation (Eigenvalue > 1, KMO = 0.800, $p = 0.000$) on the methodology questions showed sampling adequacy (Field 2009), as shown in Table 7.6. Four factors were originally identified, explaining 62% of the variance in methodology. However, the mix of loaded variables was impossible to interpret; therefore five-, three-, and two-factor solutions were tested, and the decision for a three-factor solution was taken because of interpretability (Hair et al. 2010). The factors were determined using a cut-off of 0.5 for loadings. A Haman test (Podsakoff and Organ 1986) showed that all variables loaded on their predicted factor, thus no issues with common methods bias were detected.

Table 7.5 Scale Descriptives

Measure	N	Mean	Standard Deviation	Range	Original Number of Dimensions	Scale Reliability (Alpha)	Skewness	Kurtosis
Methodology								
Comprehensive set of methodology elements	246	3.39	3.56	5.11	1	0.747	−0.629	0.094
Supplemented missing methodology elements	246	3.77	3.182	6.76	1	0.774	−1.015	2.492
Applied relevant methodology elements	246	3.98	2.63	6.16	1	0.771	−0.320	1.189
Governance								
Shareholder–stakeholder	246	2.87	4.05	4.46	2	0.741	0.419	−0.462
Behavior–outcome	246	2.97	4.75	4.51	2	0.802	−0.203	−0.617
Project success	246	3.81	3.37	4.88	5	0.923	−0.720	0.552

Table 7.6 Rotated Component Matrix For Methodology Factors

		Comprehensive Set of Methodology Elements	Supplemented Missing Methology Elements	Applied Relevant Methodology Elements
METH09	Comprehensive set of techniques	**0.809**	0.033	0.086
METH05	Comprehensive set tools	**0.783**	0.019	0.080
METH01	Comprehensive set processes	**0.762**	-0.002	0.017
METH17	Comprehensive set knowledge areas	**0.720**	-0.094	0.216
METH13	Comprehensive set cap-profiles	**0.665**	-0.002	0.163
METH06	Supplemented missing tools	-0.041	**0.769**	0.134
METH18	Supplemented missing knowledge areas	0.042	**0.713**	0.236
METH10	Supplemented missing techniques	-0.064	**0.688**	0.151
METH14	Supplemented missing cap-profiles	0.168	**0.664**	0.309
METH02	Supplemented missing processes	-0.098	**0.658**	0.080
METH11	Applied relevant techniques	0.099	0.139	**0.748**
METH07	Applied relevant tools	0.100	0.156	**0.730**
METH03	Applied relevant processes	0.057	0.125	**0.685**
METH19	Applied relevant knowledge areas	0.151	0.246	**0.631**
METH15	Applied relevant cap-profiles	0.270	0.373	**0.601**
	Cronbach's alpha	0.747	0.774	0.771
	Variance explained (5)	29.1	18.3	7.8

Extraction method: Principle component analysis.

Rotation method: Varimax with Kasier normalization

- **Governance.** Similar analyses were done for the governance questions. The data were adequate for factor analysis (normal assumptions met [KMO 0.812, $p < 0.001$]). Principle component analysis with Varimax rotation at a cut-off Eigenvalue of 1.0 for factor acceptance (Field 2009) resulted in two factors, which explained 53% of the variance: GOVCorpGov (shareholder versus stakeholder) and GOVCorp (behavior versus outcome control). Both were reliable at Cronbach alpha's of 0.743 and 0.802, respectively.

7.5 Results

7.5.1 Impact of PMM Elements on Project Success

The correlation matrix (Table 7.7) indicates positive correlations between the variables, which provides for further analysis. Hierarchical regression analysis was performed using the previously mentioned control variable and the three independent variables for a comprehensive set of methodology elements (MF01), supplemented missing methodology elements (MF02), and applied relevant methodology elements (MF03) using project success as the dependent variable, with a significance level set at 0.05. Results are shown in Table 7.8 under Step 2. All independent variables correlate significantly with project success with an R^2 of 22.3%, thus giving support for Hypothesis 1 and its subhypotheses H1.1, H1.2, and H1.3.

Table 7.7 Correlation Matrix

	ProjectSuccess REGR Factor Score 1 for Analysis 1	Project Work Experience (Years) DEM06	Comprehensive Set of Methodology Elements MF01	Supplemented Missing Methodology Elements MF02	Applied Relevant Methodology Elements MF03	GOVControl Goverance "Behavior-> Outcome Orientation"	GOVCorpGoV Corporate Goverance "Shareholder-Stakeholder Orientation"	MF01 × GOVControl	MF02 × GOVControl	MF03 × GOVControl	MF01 × GOVCorpGov	MF02 × GOVCorpGov	MF03 × GOVCorpGov
ProjectSucess REGR factor score 1 for analysis 1	1.000												
DEM06 Project work experience (Years)	-0.063	1.000											
Comprehensive set of methodology elements (MF01)	0.196****	-0.094	1.000										
Supplemented missing methodology elements (MF02)	0.168****	0.089	-0.002	1.000									
Applied revelant methodology elements (MF03)	0.385****	0.059	-0.006	0.000	1.000								
GOVControl Goverance Behavior-> Outcome Orientation	0.019	0.092	-0.157**	0.174****	-0.073	1.000							
GOVCorpGoV Corporate Goverance (Shareholder->Stakeholder Orientation)	0.270****	-0.050	0.090	-0.034	0.104*	-0.013	1.000						
MF01xGOVControl	-0.009	0.101*	0.026	0.066	-0.041	0.016	0.116*	1.000					
MF02xGOVControl	0.041	0.023	0.061	-0.357****	0.116*	0.023	0.089	-0.076	1.000				
MF03xGOVControl	0.017	0.098	-0.045	0.134*	-0.071	0.110*	-0.051	-0.040	-0.009	1.000			
MF01xGOVCorpGov	0.036	0.136*	0.176***	0.030	0.014	0.109*	-0.080	-0.233****	0.051	0.105*	1.000		
MF02xGOVCorpGov	0.139*	0.135*	0.040	0.227****	-0.012	0.110*	0.145*	0.074	-0.092	0.025	-0.121*	1.000	
MF03xGOVCorpGov	0.107*	0.043	0.018	-0.015	-0.077	-0.056	0.058	0.127*	0.009	0.137*	-0.311****	0.286****	1.000

*p≤0.05; **p≤0.01; ***p≤0.005; ****p≤0.001

Table 7.8 Hierarchical Regression with PMM as Independent Variables, Project Success as Dependent Variable, and Governance as Moderator Variable

Variables Entered	Dependent Variable Project Success (N=243)			
	Step 1	Step 2	Step 3	Step 4
Control Variable				
Project work experience	-0.063	-0.084	-0.078	-0.086
Main effect IV on DV				
MF01: Comprehensive set of methodology elements		0.191****	0.180***	0.171***
MF02:Supplemented missing methodology elements		0.176***	0.173***	0.176***
MF03:Applied revelevant methodology elements		0.391****	0.372****	0.380****
Moderators				
GOVControl (goverance control orientation) (1)			0.055	0.051
GOVCorpGov (corporate goverance orientation) (2)			0.218****	0.208****
Interaction Terms				
MF01x(1) - Control-> Behavior				-0.036
MF02x(1) - Control-> Behavior				0.030
MF03x(1) - Control-> Behavior				0.018
MF01x(2) - Shareholder -> Stakeholder				0.028
MF02x(2) - Shareholder -> Stakeholder				0.045
MF03x(2) - Shareholder -> Stakeholder				0.128*
F for regression	0.974	17.060****	14.724****	8.004****
F for change	0.974	22.335****	8,035****	1.207
R-square	0.004	0.223	0.272	0.295

Main table contains standard coefficient betas VIF <2

* $p \leq 0.05$
** $p \leq 0.01$
*** $p \leq 0.005$
**** $p \leq 0.001$

7.5.2 Moderating Effect of Governance on Relationship Between Elements of a PMM and Project Success

Following Sharma et al. (1981), a hierarchical regression analysis was carried out to test moderating influences of governance on the relationship between methodology and project success (Hypothesis H2).

The results are shown in Tables 7.5 and 7.6. Variance inflation factors (VIF) with values under 2 indicate no issues of multicolinearity among the independent variables. The control variable (DEM06) had no significant effect on the dependent variable (project success). As stated above, MF01-COMPREHENSIVE, MF03-SUPPLEMENTED, and MF03-APPLIED had a significant direct effect in Step 2 of Table 7.6, with R^2 = 22.3%.

The moderating variables GOVControl and GOVCorpGov were inserted in Step 3 (see Table 7.6). GOVCorpGOV significantly correlates with project success. The interaction effect is tested in Step 4 by inserting the product of independent variables and moderator variables. It shows that the interaction of MF03-APPLIED with GOVCorpGOV is significantly correlated with project success, thus a quasi-moderator (Sharma et al. 1981). However, the F for change in Step 4 of Table 7.8 is not significant; therefore, GOVCorpGOV can be considered as a quasi-moderator (Sharma et al. 1981).

The other governance dimension, GOVControl, does not interact with any of the independent variables but is related to MF01-COMPREHENSIVE and MF03-SUPPLEMENTED. Therefore, the visual binning was carried out for MF03-APPLIED by dividing the data into four groups to determine whether there is a significant difference between groups. The results showed no significant difference between the four bins (groups); therefore, according to Sharma, Durand, and Gur-Arie (1981), GOVcontrol is possibly an exogenous, predicator, intervening, antecedent, or a suppressor variable. This warrants further investigation.

7.5.3 Exploring the Impact of Project Governance on a PMM

In an exploratory approach, we looked at the direct impact that project governance, more specifically GOVControl (behavior versus outcome), has on the use of PMM.

GOVControl was now the independent variable and was tested against MF01-COMPREHENSIVE (a comprehensive set of methodology elements), MF03-SUPPLEMENTED (supplemented missing methodology elements), and MF03-APPLIED (applied relevant methodology elements). The results showed that the relationship between GOVControl and MF01-COMPREHENSIVE was significant ($p \leq 0.01$) with a beta of -0.163. This indicates that organizations that are more behavior/compliance oriented are more likely to have a complete set of methodology elements.

The second set of results showed that the relationship between GOVControl and MF03-SUPPLEMENTED was significant ($p < 0.005$), with a beta of 0.184. This shows that organizations that are more outcome oriented are more likely to supplement missing methodology elements, as required, than those that are more compliance oriented, who use a complete methodology.

The third set of results showed that the relationship between GOVControl and MF03-APPLIED was insignificant; therefore, GOVcontrol (behavior versus outcome) has no impact on how the methodology elements are used.

7.5.4 Other Findings

We examined project success on the basis of demographics and additional methodology data. These tests were conducted using ANOVA to examine the difference between the means of different groups selected using demographic data. There were significant differences where $p = 0.05$:

- Respondents who said they used PMMs designed for services had significantly higher project success rates than those who said PMMs were developed for products or both products and services.
- Respondents who said their PMM required a higher level of project management experience reported significantly higher project success rates.
- Respondents who said they used an international PMM were significantly more likely to report that their methodology was comprehensive.

7.6 Discussion

The three independent factors (MF01-COMPREHENSIVE, MF03-SUPPLE-MENTED, and MF03-APPLIED) represent completeness, supplementation, and application of the elements of a PMM, respectively. All three factors are significantly correlated to project success, and 22.3% of the variation in project success can be explained by applying the relevant PMM elements (MF03-APPLIED) throughout the project life cycle.

The results support the findings of White and Fortune (2002) and Shenhar et al. (2002) and show that the experience of using a PMM and the correct choice of tools, techniques, and processes are both success factors.

The results show that one of the two moderator factors, GOVCorpGov, which is the shareholder-versus-stakeholder continuum, acts as a quasi-moderator. This means that it has an indeterminate impact on the relationship between applied methodology elements (MF03-APPLIED) and project success, because in this constellation, "each of the independent variables can, in turn, be interpreted as a moderator" itself (Cohen 1988, p. 294). The other two independent variables, comprehensive set of methodology elements (MF01-COMPREHENSIVE) and supplemented methodology elements (MF03-SUPPLEMENTED), are not moderated by either of the two moderator factors.

From this point, the study turns from deductive to exploratory as we look to see if there is a direct relationship between the other moderator variable (GOVCorp) and the independent variables (MF01-COMPREHENSIVE to MF03-APPLIED). We find a significant relationship with the independent variables MF01-COMPREHENSIVE and MF03-SUPPLEMENTED. This implies that governance not only acts as a quasi-moderator (GOVCorpGov) between the applied PMM and project success, but also may influence the development or selection of the PMM, whether it is comprehensive or not. If an organization is more behavior oriented, the incumbent PMM is more likely to be enhanced over time, thereby not requiring supplementation by the project manager. However, for organizations that are more outcome oriented, there is a likelihood that the PMM will not be complete and will require supplementation by the project manager. This may be a deliberate intention to allow the project manager to tailor the PMM for the project needs.

Contingency theory within the field of project management offers insight into how to best adapt project management practices within a given environment to meet the project management goals (Donaldson 2006; Müller, Geraldi, and Turner 2012; Turner, Müller, and Dulewicz 2009; Wheelwright and Clark 1992). A PMM's completeness is contingent on governance and suggests that using contingency theory as a theoretical lens supports the premise that PMMs are impacted by context.

Additional findings suggest that project success is more correlated to stakeholder-oriented than to shareholder-oriented organizations. Project success is also associated with organizations that have comprehensive PMMs versus organizations with incomplete PMMs. The findings also show that more experienced project managers are needed to effectively apply both comprehensive PMMs and PMMs that need to be supplemented.

7.7 Conclusions

This study is the second part of a mixed-methods study that investigates the effect of governance on the relationship between a PMM and project success using a contingency theory perspective. A deductive approach validated a theoretically derived research model. The data were collected through a web-based questionnaire, with 246 respondents from six industry sectors evenly distributed across North America, Europe, and Australasia. PMM impact on project success was analyzed, including the quasi-moderating effect of governance on this relationship.

The two research questions can now be answered. For the first question, we found that there is a positive relationship between PMM and project success. Regarding project success, 22.3% of the variation is accounted for by the PMM, supporting Hypothesis 1. Sub-hypothesis H1.1 is supported, whereby having a comprehensive set of PMM elements including tools, techniques, process capability profiles, and knowledge areas (MF01-COMPREHENSIVE) is linked to project success. Also, project PMMs that are comprehensive have higher success rates than PMMs that need to be supplemented; but supplementing with PMM elements (MF03-SUPPLEMENTED) is also linked to success; therefore, Sub-hypothesis H1.2 is supported. Applying the relevant PMM elements (MF03-APPLIED) is also positively correlated with success, supporting Sub-hypothesis H1.3.

For the second research question—project governance as a moderator on the relationship between PMM and success—we observed one of the two moderating factors GOVCorpGov (shareholder–stakeholder) acting as a quasi-moderator and not as a full moderator. The role of the second proposed moderator, GOVControl (behavior–outcome), was also indeterminable, because results indicate that it can be either an exogenous, predicator, intervening, antecedent, or suppressor variable (Sharma et al. 1981). Therefore, Hypothesis 2 is only partly supported and needs further investigation.

Several researchers (Fortune and White 2006; Shenhar et al. 2002) show that it is not the use of a PMM that leads to project success; it is the experience of using a project PMM and the ability to tailor it to the context of a project that links to project success. The results of this study indicate that having a comprehensive PMM and the experience to tailor a PMM are two success factors in the context of the organizational environment. Therefore, the understanding of the organization's governance paradigm is part of the contextual positioning of how to apply the PMM.

After testing the research model, the study switched from confirmatory to exploratory research to understand whether governance has a direct impact on a project PMM. The findings suggest that project governance may also influence the selection of a PMM and how it evolves. For example, when an organization is more behavior oriented, the findings show that the organization's PMM is more likely to be comprehensive. The opposite is true for organizations that are more outcome oriented. Therefore, organizations that make a decision to develop their own PMM or adopt an international standard will have different starting points as well as different paths to whether and how their PMM evolves, depending on their governance paradigm.

7.7.1 Practical Implications

All project managers should have access to a comprehensive PMM with the experience to know which of the PMM elements to apply to any given project and, if required, supplement missing PMM elements, because collectively they accounts for 22.3% of the variation in project success.

A manager responsible for several projects who knows the governance paradigms and their implications on current and future projects may help influence, shift, or create local project governance paradigms that are more conducive to success. Organizations that have a more comprehensive PMM need experienced project managers to ensure that they achieve high success rates. By understanding the governance paradigm and the state of the evolution of the organization's PMM, a program or project portfolio manager will have insight into the project management skills and, especially, the experience necessary for a successful project outcome. When project success rates are dropping and lessons learned indicate the possibility of an unsuitable PMM, understanding the governance paradigms and the risks associated with the evolution of a PMM within each governance paradigm may provide valuable information as to the root cause of the problems.

7.7.2 Theoretical Implications

This study provides several new insights that can inform further theory development. First, PMM can now be added as a success factor to the project success literature, in that it stands for 22.3% of a project's success. This constitutes a major effect of practical significance (Cohen 1988).

Second, the study shows the importance of distinguishing between the presence of and use of PMMs. The presence of PMMs in the form of comprehensiveness (MF01-COMPREHENSIVE) or the need for supplementation (MF03-SUPPLEMENTED) carry less weigh than the application of a PMM (MF03-APPLIED) in the success equation. Accordingly, further research on project success needs to take this difference into account by being observant of the application of PMM (or other success factors) and not its mere presence. This warrants further investigation for other nonhuman-related project success factors, such as the presence versus the use of mission statements, plans, or schedules, to name a few. The results of these studies potentially change our understanding of success factors to a large extent.

Third, the selection of a project PMM and its evolution is influenced by governance. As with PMM elements, a distinction between presence and application prevails in governance. Behavior-controlled organizations prefer comprehensive PMMs, and outcome-controlled organizations prefer supplementable PMMs when being successful. However, it should be noted that application is not influenced by governance. Related theoretical implications are that governance is mainly confined to the procedural aspects, such as form selections and provisions of PMMs, but does not influence the project manager's behavior in terms of the appropriate usage thereof. Again, the project manager's work appears to be decoupled from the procedures and processes provided to him or her, which should be investigated further.

7.7.3 Further Research

Future research could provide insights into determining the effectiveness of a PMM and its elements in achieving project success by evaluating:

- Are there other moderating or mediating factors that influence the relationship between project PMM and project success?
- Which factors influence an organization to develop its own PMM or adopt a certain type of PMM, and how do these factors influence how a PMM evolves within the organization?

7.7.4 Strengths and Limitations

One of the strengths of this study is the sample and its balance between the three main regions of the world. Another strength is the targeting of professionals who are engaged in professional organizations, which led to better responses, because these respondents are interested in their profession over and above their employer's demands. This strength also comes at the cost of a limitation. The use of professional associations such as IPMA® and PMI for the distribution of the questionnaire limited the pool of respondents to only their members. A second limitation lies in the exploratory results of some of the findings, which requires further study for validation. Another limitation is that it is unclear whether the respondents' last projects were completed recently or, say, five years ago, which may influence their responses to the questionnaire.

7.7.5 Contributions to Knowledge

This chapter contributes to the understanding that the effectiveness of a PMM is determined not only by the manner in which it is applied, but in the way organizational governance paradigms influence the selection and evolution of a PMM. The effectiveness of a PMM that contributes to project success is influenced potentially by many factors wherein governance directly impacts a PMM but is only a quasi-moderating factor in the relationship between PMM and project success.

PMMs need to continually evolve by adapting to the organizational environment within the governance paradigm; otherwise, these PMMs will be misaligned with the project contexts and hence reduce their contribution to project success.

Chapter 8

The Relationship Between Project Governance and Project Success

Coauthored with Ralf Müller
BI Norwegian Business School, Norway

8.1 Introduction

This study looks at the relationship between project governance and project success from an agency theory and stewardship theory perspective. To achieve this project, governance was operationalized, respectively, as (1) the extent of shareholder versus stakeholder orientation, and (2) the extent of behavior versus outcome control, both exercised by the parent organization over its project. A cross-sectional, worldwide online survey yielded 254 usable responses. Factor and regression analyses indicate that project success correlates with increasing stakeholder orientation of the parent organization, whereas the types of control mechanisms do not correlate with project success. Results support the importance of stewardship approaches in the context of successful projects.

Forty years of research have brought a variety of new success factors (i.e., those elements that, when applied during a project's life cycle, increase the project's chances to be successful) and extend the number of success criteria (i.e., those measures applied at the end of the project to judge the project's success). Project success is hereby seen as the achievement of a particular combination of objective and subjective measures, manifested in the success criteria and measured at the end of a project (Müller and Judgev 2012).

But success rates still do not meet expectations (Judgev and Müller 2005; Lehtonen and Martinsuo 2006). Because of that, researchers have started to widen the scope of possible success factors and focus more on the structural characteristics of the project context and its impact on success. One of these factors is project governance, which has grown exponentially in popularity since 2005 (Biesenthal and Wilden 2014). This stream of literature identifies the structural characteristics needed for successful project execution (Müller and Lecoeuvre 2014). Project governance is "the use of systems, structures of authority, and processes to allocate resources and coordinate or control activity in a project" (Pinto 2014, p. 384); it coexists within the corporate governance framework with the objective of supporting projects in achieving their organizational objectives (Müller 2009). The majority of published research on project governance is conceptual, supplemented by some qualitative studies and very little quantitative evidence on the relationship between project governance and project success.

Among the few quantitative studies are Wang and Chen's (2006) assessment of the impact of governance on success in ERP projects, and Müller and Martinsuo's (2015) investigation of the role of project governance in the relationship of relational norms between project buyers and suppliers and their joint project's success. However, both studies showed an important role of governance but were confined to the IT industry. This is in contrast to general management studies, in which the link between corporate governance, management performance, and shareholder value is well researched (Amzaleg et al. 2014; Core, Holthausen, and Larcker 1999; Lazonick and O'Sullivan 2000; Maher and Andersson 2000). Because project governance is aligned with corporate governance, and good corporate governance is associated with management performance, a link between project governance and project success may be assumed. This will be addressed in the present chapter.

The purpose of this study is to investigate the relationship between project governance and project success. The aim is to understand which forms of project governance relate to project success. To achieve this, the following research question is posed:

What is the relationship between project governance and project success?

To answer this question, we first empirically test the correlation between project governance and project success. After that we discuss some of the underlying assumptions that, when met, may provide indicators for a limited causality. The unit of analysis is the relationship between project governance and project success. The study uses the governance paradigms framework from Müller and Lecoeuvre (2014), which identifies two governance dimensions:

(1) a continuum of the extent of shareholder versus stakeholder orientation (following Clarke 2004), and (2) a continuum on the level of behavior versus outcome control (following Ouchi 1980), as exercised by the project's parent organization. This allows for the contrasting views of agency and stewardship theory. Agency theory is hereby seen as a proxy in explaining behavior in more shareholder-oriented governance structures, in which contracts and process control structures are used to manage the self-serving behavior of managers for the maximization of shareholder wealth (Berle and Means 1968; Friedman 1962). Conversely, stewardship theory is taken as a lens explaining behavior in more stakeholder-oriented governance structures, in which trust and controlling by outcomes/results serve as a mechanism to govern toward the achievement of organizational goals by balancing the requirements of a diverse set of stakeholders (Davis, Schoorman and Donaldson 1997; Müller 2011).

The study is relevant for practitioners developing success-related governance structures by pointing out the success-related governance approaches, and for academics in developing contingency theories of project performance and results.

The next section reviews the literature on governance, project success, and agency and stewardship theories from which the hypotheses are derived, followed by the research methodology, results, and discussion sections. The chapter finishes with the study's conclusions and implications.

8.2 Literature Review and Hypotheses

8.2.1 Governance as a Success Factor on Projects

Building on the early success factor models by Pinto and Prescott (1988) and Pinto and Slevin (1988), which covered organizational effectiveness and technical validity, the development of success factors diversified significantly over the years. Researchers soon realized that success factors without structure, grouping, and context would result in increased project risks; therefore, success factor frameworks were introduced, such as those fostering multi-dimensionality and idiosyncrasy of factors (Baccarini 1999; Shenhar et al. 2001). Further research showed the importance of soft factors such as teamwork (Hoegl and Gemuenden 2001) or leadership styles of project managers (Turner and Müller 2005) and the shared leadership by the team (Cox, Pearce, and Perry 2003) (see Judgev and Müller [2005] a for complete review). Serra and Kunc (2014) showed the link between strategy planning and execution using benefits realization management (BRM) as a success factor.

The importance of project governance as a success factor in large-scale investment projects was empirically assessed in two qualitative case studies in South Africa. Using Delphi and nominal group techniques, the researchers found strong agreement among the interviewees that the application of governance principles affects project success (Bekker and Steyn 2008). A recent quantitative study on the impact of project management methodologies on project success in different project governance contexts used the analysis framework from Sharma, Durand, and Gur-Arie (1981). Results indicated that governance has neither a pure moderating nor a mediating role in the methodology–success relationship; thus, it indicates that governance is an antecedent variable.

This is in line with conceptual studies, which perceive governance as spanning the entire life cycle of temporary organizations, such as projects. In particular, the organization's shareholder or stakeholder orientation, as well as the organizational control structures, can be assumed to exist before individual projects are launched in these organizations. Hence, Stinchcombe's (1965) theory may apply, which suggests that, "The founding characteristics imprinted at the birth of an organization influence its subsequent behavior" (Van de Ven 2007, p. 169). Therefore, we assume, "temporal precedence of the cause [project governance] occurring before the effect [project success, measured at the end of the project]" (Van de Ven 2007, p. 169), contingent on the criteria that governance structures are set up by organizations independent of their project types—thus, governance structures are not chosen depending on the project at hand.

If this is the case, the empirical test fulfills the first of three criteria for causality, as proposed by the 19th-century philosopher John Stuart Mill and, more recently, by Andrew van den Ven (2007). The other two criteria (covariation or correlation, and absence of spurious factors) are addressed in the analysis section of this chapter. A discussion about a possible causal relationship between project governance and project success follows in the conclusion section.

8.2.2 Project Success

Historically, the understanding of project success criteria has evolved from the simplistic triple constraint concept, known as the iron triangle (time, scope, and cost), to something that encompasses many additional success criteria, such as quality, stakeholder satisfaction, and knowledge management (Atkinson 1999; Judgev and Müller 2005; Müller and Judgev 2012; Shenhar and Dvir 2007). In terms of measuring success, a variety of models for measuring project success were developed, such as the popular ones by Pinto and Prescott (1988), Shenhar et al. (2002), Hoegl and Gemünden (2001), and Turner and Müller (2006), which are all designed with different underlying assumptions.

An amalgamation of these models was done by Khan, Turner, and Maqsood (2013), who analyzed the literature on success criteria of the past 40 years. Their model for measuring success was selected for this study, because it is based on the most recent literature, which is a superset of the success criteria from the leading researchers on project success. Their model offers a balance between hard and soft factors and measures 25 success criteria variables organized in the five dimensions. The model contains the three criteria, which are typically termed the iron triangle (Dimension 1 below), plus four additional project success criteria dimensions:

1. Project efficiency
2. Organizational benefits
3. Project impact
4. Stakeholder satisfaction
5. Future potential

Table 8.1 contains the list of success criteria variables (questions).

In this chapter, project success is assessed for its correlation with project governance and then discussed as a possible dependent variable in a causal relationship.

8.2.3 Project Governance

According to Klakegg et al. (2009), it is important that governance cover all levels of the organization, starting with corporate governance flowing from the board level, to the management level responsible for execution, and down to the project level of governance. The definition of corporate governance from the Organization for Economic Co-operation and Development (OECD) is:

"Involving a set of relationships between a company's management, its board, its shareholders and other stakeholders [. . .] and should provide proper incentives for the board and management to pursue objectives that are in the interests of the company and its shareholders and should facilitate effective monitoring" (OECD 2004, p. 11).

Project-related governance is based on and aligned with corporate governance but focuses on the governance of individual projects. The Project Management Institute (PMI) defines project governance as, "an oversight function that is aligned with the organization's governance model and that encompasses the project lifecycle [and provides] a consistent method of controlling the project and ensuring its success by defining and documenting and communicating reliable, repeatable project practices" (PMI 2013a, p. 34). Whereas project governance looks at the governance of individual projects, the governance of projects looks at a group of projects, such as a program or portfolio of projects, and therefore has a broader perspective (Müller, Pemsel, and Shao 2015).

Before going into more detail on project governance, it is important to understand the history and application of management theories in the corporate governance world, because many of them apply to and are used in project governance.

Table 8.1 Success Criteria Variables (Questions)

My last project was successful in terms of:	Project Success Achieved				
	Not successful	Slightly successful	Moderately successful	Highly successful	Very highly successful
Completed according to the specification	O	O	O	O	O
Supplier satisfied	O	O	O	O	O
Enabling of other project work in future	O	O	O	O	O
Project achieved a high national profile	O	O	O	O	O
Yielded business and other benefits	O	O	O	O	O
Met client's requirement	O	O	O	O	O
Minimum disruption to organization	O	O	O	O	O
Cost effectiveness of work	O	O	O	O	O
Met planned quality standard	O	O	O	O	O
Adhered to defined procedures	O	O	O	O	O
Learned from project	O	O	O	O	O
Smooth handover of project outputs	O	O	O	O	O
Resources mobilized and used as planned	O	O	O	O	O
Improvement in organizational capability	O	O	O	O	O
Met safety standards	O	O	O	O	O
Minimum number of agreed scope changes	O	O	O	O	O
Motivated for future projects	O	O	O	O	O
Project's impacts on beneficiaries are visible	O	O	O	O	O
Project achieved its purpose	O	O	O	O	O

(continues on next page)

Table 8.1 Success Criteria Variables (Questions) (cont.)

My last project was successful in terms of:	Project Success Achieved				
	Not successful	Slightly successful	Moderately successful	Highly successful	Very highly successful
Project has good reputation	O	O	O	O	O
Finished on time	O	O	O	O	O
New understanding/knowledge gained	O	O	O	O	O
Steering group satisfaction	O	O	O	O	O
Complied with environmental regulations	O	O	O	O	O
End-user satisfaction	O	O	O	O	O
Project team satisfaction	O	O	O	O	O
Activities carried out as scheduled	O	O	O	O	O
Finished within budget	O	O	O	O	O
Sponsor satisfaction	O	O	O	O	O
End product used as planned	O	O	O	O	O
Personal financial rewards	O	O	O	O	O
Met organizational objectives	O	O	O	O	O
The project satisfies the needs of users	O	O	O	O	O
Personal nonfinancial rewards	O	O	O	O	O

Before the 1980s, corporate governance was largely in the realm of lawyers until economists became interested in how organizations make decisions (Gilson 1996). Gilson went on to say that the economists perceived a connection between organizational governance and organizational performance. From this point, researchers started to apply management theories to help understand the factors that influence corporate governance and organizational performance (Maher and Andersson 2000). The most popular theories applied to corporate governance include agency theory, stewardship theory, transaction cost economics, stakeholder theory, shareholder theory, and resource dependency theory (Yusoff and Alhaji 2012).

One of the motivations for using general management theories to ground theories in the governance of corporations was to help frame, understand, and address the issues associated with poor corporate governance (Hirschey, Kose, and Anil 2009). Since the late 1970s, the issues associated with poor corporate governance and the impact on shareholder value has been well researched across the major economies (Basu et al. 2007; Hirschey et al. 2009). Resolving issues associated with corporate governance has been shown to consistently increase shareholder gains (Gompers, Ishii, and Metrick 2003).

Agency theory, which is based on Jensen and Meckling's (1976) work takes an economic view of the shareholder and manager relationship in companies by assuming rational and self-interested actors. Agency theory has been used by researchers in traditional finance and economics—for example, accounting (Demski and Feltham 1978), economics (Spence and Zeckhauser 1971), and finance (Fama 1980)—then applied to marketing (Basu et al. 1985), political science (e.g., Mitnick 1995), organizational behavior (Eisenhardt 1985), sociology (Kaiser 2006), corporate governance (John and Senbet 1998), and project governance (Turner and Müller 2003). It posits that corporate managers (agents) may use their control over the allocation of corporate resources opportunistically in order to pursue objectives not in line with the interests of the shareholders (principals) (Jensen and Meckling 1976).

This is exemplified in the principal–agent problem, which occurs when both principal and agent act in a self-interested, utility-maximizing manner (Mitnick 1973). Davis, Schoorman and Donaldson (1997) relate this behavior to the lower levels of Maslow's (1970) hierarchy of needs. Principal agent problems arise from information asymmetry, because one party (e.g., the project manager as agent) typically has more or better information than the other (e.g., the project sponsor as principal) (Wiseman, Cuevas-Rodríguez, and Gomez-Mejia 2012). This results in a moral hazard risk which, unless mitigated, is likely to increase the agency effect (Poblete and Spulber 2012).

Popular remedies to the problem include contracts and incentives that motivate agents to act in accordance with their principals, controlled through related control structures. Corporate and project governance, when designed correctly within the context of the organization, should minimize the risks and issues associated with agency theory. Agency theory, based on Jensen and Meckling's (1976) view of principle agent models, has been criticized because it neglects to consider that the principle–agent transitions are socially embedded and therefore impacted by broader institutional contexts (Davis, Schoorman, and Donaldson 1997a; Wiseman et al. 2012). In this study, we use agency theory as a proxy to explain behavior in the shareholder-oriented and behavior-controlled governance structures.

Stewardship theory arose in response to criticism regarding the generalizability of agency theory. It takes a psychological perspective toward governance and states that the actors (managers) are stewards whose motives are aligned with the higher-level objectives of their principles rather than their own, short-term, utility-maximizing objectives (Donaldson and Davis 1991). Davis, Schoorman, and Donaldson (1997b) relate this behavior to the higher levels of Maslow's (1970) hierarchy of needs. The steward differs from the agent in that the steward is trustworthy and will make decisions in the best interests of the organization, whereas an agent needs to be incentivized and/or controlled to do this (Davis, Schoorman, and Donaldson 1997b). Stewardship theory has been criticized because it views the organization in a static way and does not account for stewards' resorting back to an agent position when their positions are threatened (Pastoriza and Ariño 2008). In the present study, we use stewardship theory as a proxy to explain behavior in the stakeholder-oriented and outcome-controlled governance structures.

Neither agency theory nor stewardship theory is more valid than the other, as each may be valid for different types of phenomena (Davis et al. 1997b). This study investigates some of these phenomena.

Both agency and stewardship theory define the relationship between actors, thus are task- or project-level theories. They are complemented by their organizational counterparts' shareholder and stakeholder theory, respectively. These are described further on this chapter.

Transaction cost economics (TCE) is an economic theory that suggests that organizations achieve the lowest transaction costs by adapting the governance structures to the nature of the transaction (Williamson 1979). Resource dependency theory suggests that managers are able to prioritize the internal and external resources needed to achieve the corporate objectives (Pfeffer and Salancik 1978). When applied, all of these theories have helped to improve corporate governance within organizations, underpinning ethical values and moral choices (Cameron et al. 2004).

In the realm of projects, two of the three elements that constitute governance are *project governance* (governance of individual projects) and *the governance of projects* (governance of a group of projects such as a program or portfolio) (Müller et al. 2015). Both elements are aligned with the Project Management Institute (PMI) definitions and governance structures of projects, programs, and portfolios (PMI 2013a, 2018).

The literature on project governance shows the diversity of governance approaches (Müller et al. 2015), covering topics such as the optimization of the management of projects (Too and Weaver 2014); the interrelationship of governance, trust, and ethics in temporary organizations (Müller and Andersen et al. 2013); risk, uncertainty, and governance in megaprojects (Sanderson 2012); governance in particular sectors, such as information technology (Weill and Ross 2004); and the normalization of deviance (Pinto 2014). Papers on governance within the realm of projects have utilized to a large extent the same management theories used in corporate governance (Biesenthal and Wilden 2014).

Quantitative studies on project governance and success were done mainly in the IT industry, where Wang and Chen (2006) used structural equation modelling to show that an equilibrium of explicit contracts, implicit contracts, reputation, and trust as governance mechanisms mediates the relationship between project hazards and project success. A study by Müller and Martinsuo (2015) showed the moderating role of project governance in the relationship of relational norms between project buyers and suppliers and their joint project's success. Thus, the number of quantitative studies is limited and industry specific. The cross-sectional study by Joslin and Müller (2015b) identified governance as a quasi-moderator, thus holding an indeterminable role in the methodology–success relationship. Complementarily, the qualitative case studies by Bekker and Steyn (2008) indicate an antecedent relationship between governance and project success. Taken together, the results show lots of variation in the role of governance in project success. This knowledge gap calls for further research.

Few publications have provided some sort of categorization system for governance and its context, such as the four governance paradigms described by Müller (2009). This model builds on two dimensions.

- The first dimension addresses the corporate-wide governance orientation by using Clarke's (2004) continuum from shareholder to stakeholder orientation of a firm.
- The second dimension addresses the control behavior exercised by the parent organization over its project, by using Ouchi's (1980) and Brown and Eisenhardt's (1997) continuum from behavior control (i.e., following the process) to outcome control (i.e., meeting pre-established expectations).

The operationalization of the paradigms was done by Müller and Lecoeuvre (2014) and allows a quantitative assessment of a project's parent organization's governance position. We chose this model for the present study because of its applicability to a wide range of projects, in an attempt to understand organizations' project governance approaches and the role of the two dimensions for project success over a wide spectrum of possible project types, industries, and geographies.

The literature on corporate governance and corporate performance shows a relationship between governance and organizational success, such that weaker governance mechanisms have greater agency problems, resulting in lower corporate performance (Hart 1995; Hirschey et al. 2009; John and Senbet 1998; Ozkan 2007); greater shareholder rights have a positive impact on corporate performance (Hirschey et al. 2009); and independent boards lead to higher corporate performance (Millstein and MacAvoy 1998). We transfer this assumption that governance timely precedes organizational success from the general management literature to the realm of projects. This follows the notions of Biesenthal and Wilden (2014), as well as Turner and Simister (2000), who see project governance as important in ensuring successful project delivery; the particular quantitative findings by Wang and Chen (2006) for governance of IT projects; and the broader findings by Joslin and Müller (2015b). Hence, we hypothesize:

Hypothesis 1. Project Governance Correlates with Project Success
The correlation between corporate governance orientation (i.e., preference for shareholder- or stakeholder-oriented governance) and project success has not been assessed in the past. A shareholder orientation of the firm is indicated when an organization prioritizes the maximization of shareholder wealth higher than the requirements of other stakeholders (Clarke 1998; Davis, Schoorman, and Donaldson 1997). Hence, when organizations take a more internal view of their raison d'être (Heblich 2010). The definition of stakeholders varies. In this chapter, we adopt Freeman's (1984) view that stakeholders are those individuals or organizations that might affect the business objectives and anyone who might be effected by its realization.

A stakeholder-oriented organization is characterized by a more external view of their *raison d'être* as an organization (Heblich 2010), which takes into account the various stakeholder groups and balances their particular requirements for the accomplishment of organizational objectives (Ansoff 1965; Clarke 1998). This is exemplified by the project management literature, which historically emphasized the importance of stakeholders in and for project success (e.g., Eskerod and Huemann 2013, plus many others). Project managers view stakeholders as the ultimate receivers of project outcome and rank their satisfaction very highly. Research shows that project managers in North America rank the importance of stakeholders highest among all success criteria, whereas project managers in other regions rank its importance consistently among the top 10 of the success criteria (Müller and Turner 2007b). Thus we hypothesize:

- **H1.1.** Stakeholder-oriented governance of projects correlates positively with project success

Similarly, the nature of the link between control orientation (behavior versus outcome) and project success is unclear from the literature. Although the literature on project management maturity models (e.g., Project Management Institute, OPM3® [PMI 2013a]) and the literature on the governance of large-scale investment projects (e.g., Klakkegg and Haavaldson [2011]) emphasize the importance of following processes for successful project implementation, other research shows a more diversified picture, such as that by Crawford et al. (2008), who showed the need for situational contingency of structures, or Turner and Müller (2004) showing that control through methodology must find the balance between being too process focused (i.e., behavior control) or too laissez-faire, because both lead to project failure. All of these studies imply a correlation between control structure and success. Given the general notion of the process orientation of project management and its maturity (PMI 2013b) and the recent popularity of process-based approaches to project management, such as Agile/Scrum (Schwaber 2004), we hypothesize:

- **H1.2.** Behavior control in project governance correlates positively with project success

Figure 8.1 shows the related research model, with the two governance dimensions on the left-hand side and project success on the right.

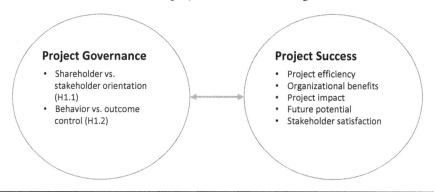

Figure 8.1   Project Governance–Project Success Research Model

8.3 Research Methodology

We followed Saunders, Lewis, and Thornhill's (2011) process for research design, which comprises seven steps: Post-positivism was used as epistemological stance, because it aims for objectivity as an ideal, but is aware of the subjectivity stemming from the subjects targeted for data collection. Post-positivism identifies trends instead of generalizations (Biedenbach and Müller 2011). A deductive approach was chosen for a robust design that includes both existing theory and new empirical evidence. A survey design was chosen to collect quantitative data in a cross-sectional manner from a wide variety of individuals in order to gain the widest coverage of the resulting theory.

8.3.1 Step 1. Questionnaire Development

Four sets of questions were included in the questionnaire. The first set included information about the last project; the next two sets covered governance paradigms and project success; and the last set collected the respondents' demographic information. The questionnaire followed the suggestions of Cooper and Schindler (2011) to ensure that the scales, criteria, and wording were consistent and clear. The project governance questions were taken from Müller and Lecoeuvre (2014). The governance paradigms were selected because they have been used successfully in several project governance–related studies before and reflect the organization's governance positioning with regard to two continuums: (1) shareholder–stakeholder and (2) behavior–outcome. The project success dimensions were based on Khan and Turner (2013). Its five dimensions—project efficiency, organizational benefits, project impact, stakeholder satisfaction, and future potential—cover the short- and long-term implications of project success. A five-point Likert scale was used, with low values representing low levels of stakeholder orientation, outcome control, and success. A pilot test was done with 10 respondents. Based on the feedback, minor wording changes were made for understandability. The pilot answers were not used in the analysis.

To avoid influences through common method bias (CMB), we followed the recommendations of Podsakoff and Organ (1986), including confirmed anonymity in the introductory text, different layouts and scales, randomizing of questions, and the conduction of Harman test for the constructs.

8.3.2 Step 2. Data Collection

A worldwide, cross-sectional questionnaire was used to collect quantitative data for generalizable results, using snowball sampling. Respondents were contacted using email with a link to the web survey. In addition, the survey details were placed on project management LinkedIn® forums. An email with the survey link was sent to PMI chapters in Switzerland, Germany, central USA, and Pakistan, asking the chapter presidents to distribute the survey link to their members. Data were collected over two weeks in April, 2014. We obtained 266 responses, of which

254 were usable for analysis. Responses came from 41 countries: 38% from North America, 24% from Europe, 22% from Australasia, and 16% from other countries. ANOVA analysis showed no difference between early and late respondents. The average respondents' work experience was 22 years, and the average project-related work experience was 15 years. Sample demographics are shown in Table 8.2.

An ANOVA test between the demographic regions showed no statistical differences ($p = 0.249$).

Project information is shown in Table 8.3. Approximately 48% of the projects were less than €1 million in cost. Of the projects, 96% were of either medium or high urgency; 42% were executed in matrix organizations and 21% in functional organizations.

8.3.3 Step 3. Analysis Methods

Analysis was carried out following the guidelines from Hair et al. (2010). Data were normally distributed (skewness and kurtosis of ±2), thus eligible for the techniques used. Eight responses were removed as outliers, because t-tests showed that the answers from these respondents were significantly different from the rest of the sample.

Analysis was done in three steps:

1. Unrotated factor analysis on each of the three constructs (governance orientation, governance control, project success) as a Harman test for possible Common Methods Bias (Podsakoff and Organ 1986)
2. Varimax rotated factor analysis (principal component analysis) with Eigenvalue of 1 to establish the factors representing each of the three constructs (Field 2009)
3. Regression analysis to test the correlation between the independent constructs (governance orientation, governance control) and the dependent construct (project success) (Van de Ven and Poole 2005)

Hence, in line with existing conventions, we tested a theoretically derived causality through correlation tests at the variable level, following Van de Ven (2007), using a variance method approach as outlined by Van de Ven and Poole (2005).

8.3.4 Validity and Reliability

Content validity was achieved by using literature-based measurement dimensions, and face validity was tested and ensured during the pilot. Construct validity was ensured through the use of published measurement dimensions (Khan et al. 2013; Müller and Lecoeuvre 2014); pilot testing of the questionnaire; and,

Table 8.2 Sample Demographics

Characteristic	N	%	Characteristic	N	%
Sector			**Gender**		
Research & development	31	12.2	Male	194	76.4
Engineering/ construction	46	18.0	Female	56	22.0
Information technology/ telecom	120	47.1	Other	1	0.4
Media/arts	9	3.5	Total	251	98.8
Relief aid	16	6.3	Missing	3	1.2
Other	29	11.4			
Total	251	98.4	**Geography—working**		
Missing	4	1.6	North America	96	37.8
			Europe	61	24.0
Position held			Australasia	56	22.0
CIO	3	1.2	Other	38	15.0
CTO	2	0.8	Total	251	98.8
Project portfolio manager	17	6.7	Missing	3	1.2
PMO	10	3.9			
Program manager	65	25.6	**Project-related experience**		
Project manager	82	32.3	1 to 5 years	36	14.6
Team member	24	9.4	6 to 10 years	63	25.6
Architect/advisor	6	2.4	11 to 15 years	53	21.5
QA/audit function	3	1.2	16 to 20 years	45	18.3
Technical stakeholder	2	0.8	20 years plus	46	18.7
Business stakeholder	4	1.6	Total	243	98.8
Other	35	13.8	Missing	3	1.2
Total	253	99.6			
Missing	1	0.4	**Work experience**		
			1 to 5 years	36	14.6
			6 to 10 years	60	24.4
			11 to 15 years	46	18.7
			16 to 20 years	49	19.9
			20 years plus	52	21.1
			Total	243	98.8
			Missing	3	1.2

Table 8.3 Project Characteristics

Characteristic	N	%	Characteristic	N	%
Duration of last project			**Urgency of last project**		
Under six months	44	17.3	Low	11	4.3
6 months to less than 1 year	67	26.4	Medium	107	42.1
1 to 2 years	76	29.9	High	135	53.1
Over 2 years	66	26.0	Total	253	99.6
Total	253	99.6	Missing	1	0.4
Missing	1	0.4			
			Last project executed in the following organizational structure		
Level of last project complexity			Projectized organization	81	31.9
Low	24	9.4	Functional organization (department)	55	21.7
Medium	117	46.1	Matrix organization	106	41.7
High	111	43.7	Other	11	4.3
Total	252	99.2	Total	253	99.6
Missing	2	0.8	Missing	1	0.4
Value of last project					
Under 500,000 (Euro)	85	33.5			
500,000 to 999,999	37	14.6			
1,000,000 to 4,999,999	61	24.0			
5,000,000 to 50,000,000	43	16.9			
Over 50,000,000	27	10.6			
Total	253	99.6			
Missing	1	0.4			

quantitatively, through unrotated factor analyses. Convergent and discriminant validity were tested and achieved through item-to-item and item-to-total correlations above 0.3 and 0.5, respectively. Reliability can be assumed with all constructs showing Cronbach Alpha values higher than 0.70 (Hair et al. 2010).

No indication for possible common method bias was found, as a Harman test showed that all questionnaire items loaded on their respective factor (Podsakoff and Organ 1986).

8.4 Data Analysis and Results

Varimax rotated factor analysis was used to establish the three constructs. Here a KMO of 0.8 ($p < 0.001$) indicated the data's appropriateness for this analysis (Hair et al. 2010). All questionnaire items loaded on their respective factor and were of acceptable reliability (Cronbach Alpha), see Table 8.4 (on next page).

- **Project success.** The factor on project success comprises five subdimensions (project efficiency, organizational benefits, project impact, future potential, and stakeholder satisfaction). A second-order factor analysis combined these subdimensions into a single factor for project success (KMO 0.930, $p < 0.001$) with high reliability (Cronbach's alpha 0.923).
- **Project governance.** The questions on governance loaded on the two respective subdimensions (KMO 0.812, $p < 0.001$), which explained 53% of the variance in GOVorientation (shareholder–stakeholder) and GOVcontrol (behavior–outcome). Both were reliable with Cronbach's of 0.743 and 0.802, respectively. GOVorientation (shareholder–stakeholder) comprised the upper five questions shown in Table 8.5 (i.e., the governance questionnaire). GOVcontrol (behavior–outcome) comprised the lower five questions in Table 8.5.

8.4.1 Correlation Between Project Governance on Project Success

Table 8.6 shows the correlation matrix of the variables.

Multi-variate regression analysis was done with project success as the dependent variable and GOVorientation (shareholder–stakeholder) and GOVControl (behavior–outcome) as independent variables. Table 8.7 shows the coefficient table.

A significant model ($p < 0.000$) with an R^2 of 0.063 and no issue with multi-colinearity (VIF < 2) was obtained. The correlation between GOVorientation (shareholder–stakeholder) and project success was positive and significant ($p < 0.001$, beta = 0.250), supporting H1.1. This constitutes a small, but significant, effect size, also known as *practical significance* (Cohen 1988). However, GOVControl (behavior–outcome) was not significantly correlated to project success at $p = 0.05$, which rejects H1.2.

The hypothesized correlation between project governance and project success (H1.1) is supported through the significant correlation. Furthermore, tests with the various demographic variables as control variables indicated no presence of spurious variables. That fulfills the two other criteria that need to be met before commencing a discussion on possible causality (Van de Ven 2007).

(text continues on page 163)

Table 8.4 Scale Descriptives

Measure	N	Mean	Standard Deviation	Range	Original Number of Dimensions	Scale Reliability (Alpha)	Skewness	Kurtosis
Governance								
Shareholder–stakeholder	246	2.87	4.05	4.46	2	0.741	0.419	-0.462
Behavior–outcome	246	2.98	4.75	4.51	2	0.802	-0.203	-0.617
Project success—dimensions (SA01 to SA05)	246	3.81	3.37	4.88	5	0.923	-0.720	0.552
SA01 Project efficiency	246	3.56	0.78	3.63	1	0.913	-0.471	-0.061
SA02 Organizational benefits	246	3.82	0.71	3.20	1	0.898	-0.563	0.062
SA03 Project Impact	246	3.95	0.79	3.75	1	0.899	-0.985	1.192
SA04 Future Potential	246	3.71	0.84	3.75	1	0.911	-0.743	0.372
SA05 Stakeholder Satisfaction	246	4.01	0.73	3.50	1	0.906	-0.774	0.649

Table 8.5 The Governance Questionnaire

In my organization....								
... decisions are made in the best interest of the shareholders and owners of the organization and their return on investment (RoI)	O	O	O	O	O	O	O	... decisions are made in the best interest of the wider stakeholder community (incl. shareholder, employees, local communities, etc.)
... the remuneration system includes stock options for employees and similar incentives that foster shareholder RoI thinking	O	O	O	O	O	O	O	... the remuneration system provides incentives for community, environmental, humanitarian, or other non-profit activities outside and/or inside the organization
... prevails an image that profitability determines the legitimacy of actions (including projects)	O	O	O	O	O	O	O	... prevails an image that wider social and ethical interests determine the legitimacy of actions (including projects)
... I am sometimes asked to sacrifice stakeholder satisfaction for the achievement of financial objectives	O	O	O	O	O	O	O	... I am sometimes asked to sacrifice the achievement of financial objectives for improvement of stakeholder satisfaction
... the long-term objective is to maximize value for the owners of the organization	O	O	O	O	O	O	O	... the long-term objective is to maximize value for society

(continues on next page)

Table 8.5 The Governance Questionnaire (cont.)

The management philosophy in my organization favors . . .						
. . . a strong emphasis on always getting personnel to follow the formally laid down procedures	O	O	O	O	O	. . . a strong emphasis on getting things done, even if it means disregarding formal procedures
. . . tight formal control of most operations by means of sophisticated control and information systems	O	O	O	O	O	. . . loose, informal control; heavy dependence on informal relationships and the norm of cooperation for getting things done
. . . a strong emphasis on getting personnel to adhere closely to formal job descriptions	O	O	O	O	O	. . . a strong emphasis to let the requirements of the situation and the individual's personality define proper on-job behavior
. . . support institutions (such as a PMO) should ensure compliance with the organization's project management methodology	O	O	O	O	O	. . . support institutions (such as a PMO) should collect performance data in order to identify skills and knowledge gaps
. . . prioritization of methodology compliance over people's own experiences in doing their work	O	O	O	O	O	. . . prioritization of people's own experiences in doing their work over methodology compliance

Table 8.6 Correlation Matrix

	ProjectSucess (5 combined dimensions) DV	SA01 Project Efficiency (Dimension 1) DV	SA02 Organizational Benefits (Dimension 2) DV	SA03 Project Impact (Dimension 3) DV	SA04 Future Potential (Dimension 4) DV	SA05 Stakeholder satisfaction (Dimension 5) DV	GOVControl Goverance 'Behavior--> Outcome Orientation' IV	GOVorientation (Shareholder->Stakeholder) Orientation IV
ProjectSucess (5 combined dimensions - DV)	**1.000**							
SA01 Project Efficiency (Dimension 1) - DV	0.845****	1.000						
SA02 Organizational Benefits (Dimension 2) - DV	0.902****	**0.689******	1.000					
SA03 Project Impact(Dimension 3) - DV	0.899****	0.717****	0.763****	**1.000**				
SA04 Future Potential (Dimension 4) -	0.861****	**0.627******	**0.778******	0.696****	1.000			
SA05 Stakeholder satisfaction	**0.873******	0.680****	0.716****	0.755****	0.676****	1.000		
GOVControl 'Behavior--> Outcome Orientation' IV	0.007	0.006	0.015	0.015	-0.011	-0.003	1.000	
GOVorientation (Shareholder->Stakeholder) Orientation IV	0.250****	0.237****	0.236****	0.204****	0.258****	0.162**	0.000	1.000

$*p \leq 0.05; **p \leq 0.01; ***p \leq 0.005; ****p \leq 0.001$

Table 8.7 Coefficients Table

Coefficients

Model	Unstandardized Coefficients		Standardized Coefficients	t	Sig.	Correlations			Collinearity Statistics	
	B	Std. Error	Beta			Zero-order	Partial	Part	Tolerance	VIF
(Constant)	5,115E-16	,062		,000	1,000					
GOVControl governance 'control -> behavior' orientation	,007	,062	,007	,111	,912	,007	,007	,007	1,000	1,000
GOVCorpGov corporate governance (share->stake holder) orientation	,250	,062	,250	4,024	,000	,250	,250	,250	1,000	1,000

Dependent Variable: ProjectSucess REGR factor score 1 for analysis 1

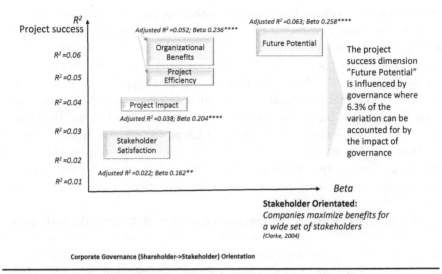

Figure 8.2   Influence of GOVorientation

Subsequently, an exploratory analysis was done to analyze the nature of the relationship between GOVorienation and project success. The five dimensions of project success—project efficiency, organizational benefits, project impact, future potential, and stakeholder satisfaction—were regressed as dependent variables against GOVorientation as the independent variable. The results showed that GOVorientation (shareholder–stakeholder) was positively and significantly correlated with all five success dimensions. The details are shown in Figure 8.2.

The success dimension future potential has the strongest correlation with GOVorientation (Adjusted R^2 = 0.063; Beta 0.258****), whereas stakeholder satisfaction has the weakest correlation of the five dimensions with an adjusted R^2 = 0.022; Beta 0.162**.

8.5 Discussion

The two independent constructs—GOVorientation (shareholder–stakeholder) and GovControl (behavior–outcome)—were tested on their relationship with project success. Only GOVorientation (shareholder–stakeholder) is significantly correlated to project success, where 6.3% of the variation in project success can be explained by the governance position along the shareholder–stakeholder continuum. With a beta of 0.25 (p < 0.001), an increase in stakeholder orientation correlates with an increase in project success. The results are consistent with the findings of Joslin and Müller (2015b), who showed that organizations that are more stakeholder oriented have greater chances of success in applying the relevant methodology elements or parts in their projects. The results also support findings in IT projects, in which governance takes a mediating role between project hazards and success by directly influencing project success (Wang and Chen 2006). Finally, the results give quantitative support to the qualitative study by Bekker and Steyn (2008), whose interviewees predicted such a relationship. Surprisingly, the second independent construct, GovControl (behavior–outcome) orientation, does not correlate with project success. In line with the literature cited above, this is indicative of a situational contingency of control structures, in that organizations in which governance is more behavior–control oriented do not necessarily achieve higher rates of project success than organizations that are outcome oriented.

The finding challenges the governance aspects of frameworks such as the Carnegie Mellon University's Capability Maturity Model Integration (CMM Integration[SM]), or the governance process/outcome orientation behind the Project Management Institute's Organizational Project Management Maturity Model (OPM3®) (PMI 2013a), wherein the premise is that a stronger process control leads to better organizational results. Along this line, Yazici (2009) showed that maturity models have helped to improve project success on a repeatable basis only in certain organizational cultures. Using the competing values framework (Cameron and Quinn 2006), Yazici demonstrated that the clan culture, which represents the importance of stakeholder participation, cohesion, shared values, and commitment, is the model most linked to project success. This underpins stewardship theory, which proposes that the behavior of individuals in organizations is aligned and supportive to the organizational and collectivistic goals instead of individualistic and self-serving goals. Project managers (agents) are tasked with complex projects and need to get things done; therefore, flexibility and trust is required from their principal (Turner and Müller 2004).

Referring to Figure 8.2, the success dimension future potential that relates to enabling, motivating, and improving an organization's capability to undertake future project work is the dimension most strongly correlated with the governance orientation. This is supported by the notion that stakeholder orientation is underpinned by balancing the requirements of several stakeholder groups simultaneously, instead of shareholders only (such as the shareholders of a project delivery organization), which is the basis for long-lasting business relationships, as outlined in Donaldson and Preston's (1995, p.67) thesis that "corporations practicing stakeholder management will, other things being equal, be relatively successful in conventional performance terms (profitability, stability, growth, etc.)."

This also applies to the other four success dimensions—namely, organizational benefits, project efficiency, project impact, and stakeholder satisfaction, which are all part of conventional performance measures at both project and corporate level. In summary, all five project success dimensions are positively correlated in varying degrees by a stakeholder orientation in project governance.

8.6 Conclusions

This study empirically investigated the relationship of project governance and project success. A deductive approach tested a theoretically derived research model. Two theoretical lenses were used in the study: agency theory and stewardship theory. The data were collected through a web-based questionnaire with 246 respondents from 11 industries evenly distributed across North America, Europe, and Australasia. The research question can now be answered: Project governance has a small, but significant correlation with project success.

Hypothesis 1 is partly supported as one of the two governance dimensions that correlates project success. H1.1 is supported because the stakeholder orientation in governance correlates positively with project success. Approximately 6.3% of the variation of project success correlates with the stakeholder orientation of the governance structure. The section on theoretical implications below outlines some of the contingencies under which this correlation might be assumed to become causal in nature—that is, the underlying assumptions that need to be met and held constant for assuming that success is to some extent dependent on project governance. H1.2 is not supported, because the governance control orientation (behavior–outcome) does not correlate with project success.

This study's results indicate the importance of understanding the governance orientation of the organization governing projects and the potential enabling effect of a stakeholder orientation in project governance for project success. Yazici (2009) found that culture impacts project success; organizations that are more stakeholder participative, cohesive, and have shared values and commitment are most likely to achieve project success. Stakeholder-oriented organizations that have shared values suggest that stewardship relationships are in place. However, this can only occur when the necessary situational factors and structures are present, including individuals with the appropriate psychological profiles (Toivonen and Toivonen 2014). When there is a change of culture in the organization resulting from external pressures—for example, a push for short-term benefits, where management trust turns into excessive control—will lead to agency tendencies (Clases, Bachmann, and Wehner 2003). Determining the appropriate governance structures should take into consideration the implications resulting from agency and stewardship perspectives toward governance and the implications stated below.

8.6.1 Practical Implications

Managers influencing the design of project governance should be aware of the importance of a stakeholder orientation for project success. This should be included in training programs for these managers, at the industry as well as the academic level. This includes courses in (project) governance, mid- and higher-level management trainings, organizational design courses, etc.

Simultaneously, managers should be aware that control structures that foster behavior or outcome control do not correlate with impact project success on a global basis, but may do so in the particular circumstances of their projects.

Recruitment managers should understand the personality traits of project managers and their governors to ensure that their personalities are aligned to a stewardship role within the project governance environment.

Project managers should understand their organization's governance procedures and work with the authority that defines project governance procedures to tailor the procedures to the project environment and/or project type.

8.6.2 Theoretical Implications

In this section, we discuss the conditions for assuming a causal relationship between project governance (as cause—i.e., independent variable) and project success (as effect—i.e., dependent variable). Throughout the chapter we have listed the most often used "conditions researcher[s] look for in testing cause and effect relationships," as stated, for example, in Hair et al. (2003, p. 64) and supported by Van de Ven (2005) and John Stuart Mill:

1. **Time sequence.** The cause must occur before the effect.
2. **Covariance.** A change in the hypothesized independent variable is associated with a change in the dependent variable.
3. **Non-spurious associations.** The relationship is not due to other variables that may affect cause and effect.
4. **Theoretical support.** A logical explanation for the relationship.

The cross-sectional design has supported testing Conditions 2 and 3. Thus, we have shown that covariance exists (Condition 2) in the form of a significant correlation between the variables. We have also tested for non-spurious associations (Condition 3) by controlling several variables in the regressions. However, the cross-sectional design of the research does not allow us to test whether the cause (the existence of a governance structure) precedes the effect (project success). To assume causality, the governance structure must be established before a project is chosen. This may be the case in organizations that do not adjust their governance structures to the type and size of the projects they take on. However, in many cases, it is likely that governance structures are chosen based on the project type. The latter is supported, among others, by transaction costs economics (Williamson 1979), which claims that governance structures are established contingent on the specificity of the transaction's (i.e., the project's) outcome, its general risk, and its frequency. This view contrasts with, for example, Bekker and Steyn's (2008) qualitative (i.e., opinion-based) findings that project governance impacts project success. To that end, we do not find clear evidence for Condition 1.

In terms of testing for Condition 4, we have shown in the literature review section that published research on governance often assumes and tests for a causal relationship between governance and organizational success. The importance of stakeholder management in projects echoes the results that stakeholder orientation in governance correlates with better project results. However, in line with the paragraph above, we cannot rule out alternative explanations. These include the possibility that projects with higher risk levels are governed more rigorously than those with lower risk levels—that is, higher risk such as those with more shareholder orientation and from the agency theory perspective—in contrast to less rigorous and stewardship-driven governance for lower-risk projects. Support for this is indicated by Klakegg et al. (2008) and Müller and Lecoeuvre (2014), who showed that larger projects, such as public investment projects, are subject to stricter governance approaches than smaller projects. If lower-risk projects fail less often than higher-risk projects, the correlation between stakeholder orientation and project success is impacted by the spurious variable project risk, which was not tested in this study.

Hence, we cannot claim causality. A limited causality may be assumed when the following conditions exist: (1) the governance structure exists before a project is chosen; (2) the governance structure is independent of the project type, size, and risk; and (3) the governance structure does not change during the course of the project. This should be tested through future research.

Stewardship theory, which is operationalized in this study as the combination of stakeholder-oriented governance and outcome-oriented control in project governance, was shown to be an appropriate lens for assessing project governance. The findings provide evidence for a generalization to a theory (in the sense of Yin 2009) with respect to stewardship theory's applicability for project settings and a generalization to the wider population of projects and their governance. Stewardship theory and stakeholder theory are recommended as theoretical lenses for the development and implementation of project governance structures.

Simultaneously, the study shows some of the limitations of existing agency theory approaches, especially shareholder theory–driven approaches to governance. Agency theory was operationalized in this study as the combination of shareholder orientation and behavior control, which relies merely on unilateral return on investment thinking and control as the governance principle. The study's results show that these approaches are limited in their likelihood to predict project results.

The implications for developing a broader theory of project governance is that a shareholder or stakeholder orientation in project governance is required to be implemented in a way that allows it to flourish within a corporate governance structure which may or may not be supportive of it, without creating conflicts or friction points. To do that, further research is required to identify the interfaces between project and corporate governance, which can then be used to adapt the two levels of governance to each other.

8.6.3 Strengths and Limitations

The strength of the study includes the use of tested and validated measurement constructs. Another strength lies in the well-balanced sample covering the three main regions of the world, and respondents who are professionals engaged in professional organizations, which led to better responses, because these individuals are interested in their profession over and above their employer's demands.

The use of professional associations such as IPMA® and PMI for distribution of the questionnaire limited the pool of respondents to only their members. A further limitation of the study was the use of one particular governance model. Other governance models should be used for similar analyses to get a more holistic picture of the relationship between governance and success.

8.6.4 Further Research

In addition to the suggestions above, we suggest that future research should address the nature of the link between project success dimensions and project governance, as well as possible moderator or mediator effects that influence this relationship.

Further qualitative and quantitative research is needed to investigate whether project governance orientation structures optimized for project success can exist and thrive throughout an organization and under what conditions, even though the main organization's governance orientation may be different.

Process studies such as those suggested by Langley et al. (2013) are recommended in order to understand the temporal nature of the elements of project governance, their relationships, and the variations across project life-cycle stages.

Moreover, future research should investigate the impact of the governance paradigms on the governance of projects at the program and project portfolio levels, and if different, provide insights as to which paradigm(s) are the most correlated to program and project portfolio success.

The study's contribution to knowledge lies in its clarification of a correlation between different project governance approaches and project success. To that end, we have provided the ground for further studies on causality and its direction in order to investigate the role of governance as a success factor in projects.

Chapter 9

Using Philosophical and Methodological Triangulation to Identify Interesting Phenomena

Coauthored with Ralf Müller
BI Norwegian Business School, Norway

9.1 Introduction

The scarcity of accepted research designs within each research philosophy paradigm limits the variance of research approaches, which reduces the chances to identify real new phenomena. We propose that researchers use triangulation of alternative research philosophies to identify interesting new phenomena, provide alternative perspectives to complex problems, and gain a richer and more holistic understanding of complex project management problems. Philosophical triangulation extends methodological triangulation into the realm of ontology and provides for more comprehensive understanding, in that it resembles a more realistic view toward social science phenomena, which, by their nature, appear differently to people, and thus are seen from different ontological perspectives simultaneously. Three related studies are used to exemplify the approach, whereby the results of two sets of empirical data (qualitative and quantitative) are discussed in different philosophical contexts. Implications for scholars include more practice-oriented research perspectives in line with the projects-as-practice stream by extending existing benefits from methodological triangulation into philosophical triangulation in order to identify and understand complex phenomena.

Research in project management has been criticized for its lack of relevance for practitioners (Blomquist et al. 2010; Sahlin-Andersson and Söderholm 2002). As a result, several streams of literature developed in support of more practice-oriented approaches to research, which is manifested in new perspectives toward project management, rethinking papers, and broader concepts (Svejvig and Andersen 2014). However, this trend is not matched by a development in research designs (Müller and Söderlund 2015). Blomquist et al. (2010) suggest increasing a practical relevance approach to project management research by first understanding what people do within the context of projects before such projects are investigated.

Researchers following these and other related suggestions are immediately confronted with the fact that research is typically done from a narrow theoretical perspective, involving one or, at most, two theoretical lenses toward the phenomenon under study, whereas practitioners hold a multitude of perspectives simultaneously. Examples include the popular governance theories in management, such as *agency theory* (Jensen and Meckling 1976), which assumes a *homo economicus,* motivated by the lower levels of Maslow's hierarchy of needs (Maslow 1970). With its economic focus, this theory fails to explain, for example, altruistic, loyal, or other behavior related to the higher levels of Maslow's theory. This is done through *stewardship theory* (Davis, Schoorman, and Donaldson 1997), a complementary theory to agency theory.

Whereas most of the research is done from either an agency or a stewardship perspective, the practitioner in a governance situation does not know which theory to apply at what point in time; thus, the practitioner does not know which theory to use to develop a governance system in terms of what to expect regardless of the theoretical lens used and what to expect when using either one of the two perspectives.

A comprehensive understanding of phenomena arises from a researcher's simultaneous look at a phenomenon from both perspectives. This is typically done using mixed-methods studies, an approach increasingly popular in recent years. Cameron and Molina-Azorin (2011) estimate that about 14% of business and management studies use mixed methods. However, in project management research, this number is as small as 1.5% (Cameron, Sankaran, and Scales 2015), indicating that the vast majority of researchers use a singular paradigm to understand the phenomenon under study, which does not align with the practitioners' perspectives. This single-paradigm approach either produces results of questionable relevance for practice or fails to identify phenomena of practical relevance.

Moreover, within a singular paradigm, the number of accepted research designs is limited. This leads to repetitive use of similar research designs, which then leads to almost predictable research results (Müller, Sankaran, and Drouin 2013; Williams and Vogt 2011).

In this chapter, we argue that the application of several philosophical perspectives, which include the use of mixed-methods studies, provides for more practice-relevant identification and understanding of phenomena. Applying several perspectives simultaneously comes closer to the practitioners' reality and thereby creates more realistic situations for researchers. We further argue that more than two perspectives toward the same phenomenon will provide a more comprehensive identification of the phenomenon per se, its context, and its scope. This approach extends over and above methodological triangulation into the realm of ontologies and uses philosophical triangulation (Bechara and Van de Ven 2011), which makes use of methodological triangulation at the epistemological level.

The purpose of this chapter is to show that philosophical triangulation can help to identify phenomena while gaining a deeper and richer understanding of the phenomena using a natural-science comparative that otherwise cannot be explained within or across the respective research paradigm(s). Using this approach in project management research may lead to previously unobserved phenomena within a particular paradigm that is discovered but cannot be explained within the context of the paradigm.

This leads to the research question:

Is it possible to use philosophical triangulation to identify interesting phenomena, as well as to provide alternative perspectives?

The benefits of this study are to break free of the constraints of a single paradigm and its accepted methods, therefore (1) allowing the researcher to identify phenomena that may not be identified using a single paradigm, as well as (2) providing an alternative perspective through the use of a natural-science comparative.

The chapter continues with a review of related literature on triangulation and continues with the description of the multidimensional approach for philosophical triangulation. This is followed by the application of the approach by triangulating three distinct philosophical perspectives, which provides for new insights and new phenomena. The chapter finishes with a discussion and a conclusion.

9.2 Literature Review and Hypotheses

We set out to address the question of relevant research for practitioners as a "knowledge production problem" in the sense of Van de Ven (2007), created through an unengaged process of inquiry, in which researchers deal with single theoretical models for addressing the research problem or question. We build on Van de Ven's suggestions for scholars being engaged with practitioners and other stakeholders and suggest adding philosophical multiplicity to the research design. This is required because of the limitations stemming from the use of singular research paradigms.

9.2.1 Limitations of Current Research Approaches

The research designs being accepted within a research paradigm shape the nature of the studies and impact the research results. Too often researchers adjust the research questions to the methods they are familiar with, instead of adjusting the research design to the questions (Williams and Vogt 2011). This reduces the variance in research designs, which in turn leads to repetitive and narrowly designed studies with often predictable results (Müller et al. 2013). A consequence of this approach is the risk of carrying out research that not only provides predictable results but also finds fewer or potentially less interesting phenomena.

As phenomena are described within theories, the theories will also be considered more or less interesting. According to Davis (1971), interesting theories—hence interesting phenomena—are those that deny certain assumptions of their audience, while noninteresting theories are those that affirm certain assumptions of their audience. This implies that interesting theories (or phenomena) are more impactful than less interesting theories (or phenomena).

A consequence of less interesting theories and phenomena is that these are often forgotten and rarely cited. Perhaps a more concerning aspect is that, if the trend in current research theories falls into the category of "less interesting," this signals the need for new and alternative research approaches. A study by Turner, Pinto, and Bredillet (2011) showed that the number of conceptual papers and new techniques dropped by 10–25% between 1997 and 2007 in two of the three main research journals in project management. This indicates a decline in the discovery of new phenomena. This decline appears at a time when the variety of research designs is stagnating (Biedenbach and Müller 2011; Müller and Söderlund 2015).

We propose a link between these two observations: (1) repetitive use of similar research designs limits the researchers' perspectives, and (2) the chance to identify and understand real new phenomena. This chapter addresses this shortcoming through philosophical triangulation (Bechara and Van de Ven 2011). This approach allows us to leverage the strengths of different research designs following Flyvbjerg's notion (2001, p. 53) of "where natural science is weak, social science is strong, and vice versa," by combining natural and social science methods to identify the many facets of a phenomenon, thereby coming closest to the view that a practitioner has of a phenomenon.

9.2.2 Triangulation

Many researchers strive to provide rich data that are unbiased and can be understood with a comfortable degree of assurance (Breitmayer, Ayres, and Knafl 1993; Jick 1979). One way is to decrease biases, increase validity and strength of the study, and provide multiple perspectives by using methods that involve triangulation (Denzin 1970). The term *triangulation* is a military term used for surveying, which has been used as a metaphor in social science (Smith 1975).

The concept of triangulation is that when you need to locate your position on a map, a single landmark can only provide the information that you are situated somewhere along a line in a particular direction from the landmark. With two landmarks, however, your exact position can be pinpointed by taking bearings on both landmarks; you are at the point where the two lines cross.

In social science research, the analogy is that when one relies on a single piece of data, there is the danger that an undetected error in the data production, bias, or methodology process may render the analysis incorrect. Therefore, to triangulate in social science, the combination of two or more data sources, investigators, methodological approaches, theoretical perspectives (Denzin 1970), or analytical methods (Kimchi et al. 1991) about a measurement (Campbell, Schwartz, and Sechrest 1966) is used to find out "if a hypothesis can survive the confrontation with a series of complementary triangulations of testing" (Campbell and Fiske 1959, p. 82).

In the social sciences, the use of triangulation can be traced back to Campbell and Fiske (1959) and then developed further by Denzin (1970), wherein multiple triangulation was first introduced—for example, two data sources along with two investigators. So by triangulating, researchers can hope to overcome the weakness or intrinsic biases and problems that come from a single method, a single observer, a single data source, and single-theory studies.

9.2.3 Types of Triangulation

According to Denzin (1970), there are four types of triangulation:

First is a *data* triangulation, which involves using multiple sources of data during a study. Data sources can vary based on the times when the data were collected, the location of collection, and the person/people who obtained the data (Denzin 1970; Mitchell 1986).

The second type of triangulation is the *investigator* triangulation, in which more than one investigator is used in a study to engage in observations, interviews, coding, or analysis of participants' responses. Using multiple investigators reduces the potential bias inherent in employing only one investigator or analyst by allowing for data consistency through auditing. Independent confirmation of data among investigators lends greater credibility to the observations (Denzin 1970).

The third type is *methodological* triangulation, which is also known as multi-method, mixed-method, or methods triangulation (Greene and Caracelli 1997). This is the most commonly used type of triangulation (Hastings and Salkind 2013) which uses multiple methods to study a research problem. Qualitative and quantitative methods may be used simultaneously (e.g., conducting a case study while distributing a questionnaire). Methodological triangulation can be classified into two types: *within-method* triangulation and *between-* or *across-method* triangulation. Within-method triangulation uses at least two data-collection procedures from the same design approach (Denzin 1970). For *quantitative* approaches, the procedures could consist of a survey questionnaire using existing information from a database. In *qualitative* approaches, nonparticipant observations could be combined with focus group interviews. These methods are either qualitative or quantitative, but not both. Researchers using between- or across-method triangulation employ both qualitative and quantitative data collection methods in the same study (Denzin 1970; Tashakkori and Teddlie 2009).

The fourth type is *theory* triangulation, which provides multiple theoretical perspectives either in conducting the research or in interpreting the data. Multiple theoretical perspectives, such as from a marketing theory and a leadership theory, can help to rule out competing hypotheses and reduce the risk of premature acceptance of plausible explanations while increasing confidence in developing concepts or constructs in theory development (Banik 1993).

Another type of triangulation is described by Bechara and Van de Ven (2011) as *philosophical* triangulation. In this case, triangulation is performed from alternative philosophical perspectives. This provides for a richer and more holistic understanding of complex managerial and organizational problems, because each philosophy sheds light on a different aspect or facet of the phenomenon. It reveals the interdependence among various dimensions of the phenomenon and overcomes instability risks stemming from a singular perspective. Moreover, it adds to (a) reliability through converging information from different methods, and (b) validity through discussion of the divergent information from different methods. This method of triangulation is proposed in this chapter.

The literature on triangulation, regardless of what type of triangulation is used, focuses mostly on reducing bias and increasing the validity of expected phenomena. This is typical for the usage of the first three types of triangulation—that is, data, observer, and methodological triangulation. The literature on philosophical triangulation helps to improve validity indirectly by providing alternative explanations for a phenomenon (Mitchell 1986). This is a point of departure from the first three triangulation methods. Theoretical and philosophical triangulation help to create different perspectives that can help to support or disprove competing hypothesis, but—more importantly—prevent premature acceptance of plausible rationale, creating confidence in developing concepts of theory development (Banik 1993).

The literature on theoretical and philosophical triangulation only covers how alternative perspectives can help provide confidence in the accepted hypothesis. There is a research gap for using philosophical triangulation to discover interesting phenomena that may not be part of the original research question. This gap may be due to the use of the triangulation metaphor, which only uses two points to triangulate the third point.

This is explained in Figure 9.1, in which phenomenon "E" has been triangulated by using alternative philosophical paradigms. But what if there are two additional phenomena, "A" and "B"? Using the philosophical triangulation as shown in Figure 9.1, it shows that phenomenon "A" was observed—for example, using a critical realist ontology (and qualitative epistemology QUAL), which cannot be explained from the positivist perspective (natural-science comparative–conceptual). Likewise, if phenomenon "B" was observed from a positivist perspective of the conceptual study, it cannot be explained from a critical realism perspective and, therefore, is methodological specific.

This creates a dilemma for the researcher as to whether the unexplained aspects relate to the same or different phenomena. There is a risk that researchers stop at that point and fail to further test the results, leading to merely a reporting of findings, and probably not a contribution to the development of new theory—and thus, a reduction in the richness and depth of information that could have been used to help to identify and understand a new phenomenon E. If A and B were also proved to be phenomena, they might also be related to phenomenon E and, therefore, would influence the identification of and understanding of E and a theory development covering A, B, and E.

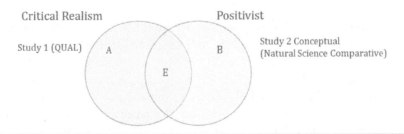

Figure 9.1   Two-Point Philosophical Triangulation

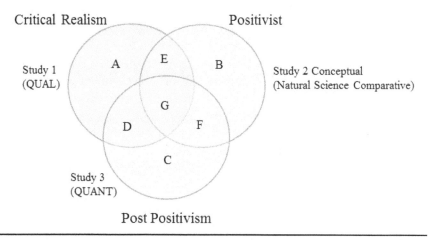

Figure 9.2   Three-Point Philosophical Triangulation

When there are three points of observation (i.e., three alterative perspectives or philosophical positions), it is possible to discover new and interesting phenomena that may not have been observed before through a two-point triangulation.

Referring to Figure 9.2, three philosophies (or ontological positions) are used, which allows for four sets of triangulation.

- **Intersect A.** Observed phenomena from the main QUAL research, which cannot be explained from a natural science perspective or from the QUANT main study research; therefore, it is methodology specific.
- **Intersect B.** Observed phenomena from the conceptual study, which cannot be explained from any other perspective; therefore, it is methodology specific.
- **Intersect C.** Observed phenomena from the main QUANT research, which can be explained neither from a natural science perspective nor from the QUAL research of the main study; therefore, it is methodology specific.
- **Intersect D.** Observed phenomena from the main studies for QUAL and QUANT research giving philosophical triangulation but cannot be explained from a natural science perspective; therefore, this part is methodology specific.
- **Intersect E.** Observed phenomena from the main QUAL study, which can also be explained from a natural science perspective giving philosophical triangulation but cannot be explained in the QUANT research; therefore, this part is methodology specific.

- **Intersect F.** Observed phenomena from the main QUANT, which can be explained from a natural science perspective, giving philosophical triangulation, but cannot be explained from the QUAL research; therefore, this part is methodology specific.
- **Intersect G.** Observed phenomena, which can be explained from a natural science and both QUAL and QUANT perspectives; therefore, giving a full philosophical triangulation.

Therefore, using a tri-philosophical triangulation wherein, perhaps, the metaphor triangulation should be taken from the perspective of what is being used to observe rather than what is being observed. Here it is possible to triangulate an additional set of three phenomena—E, D, and F—which will allow for a deeper and richer identification and understanding of the phenomenon's face G with E, D, and F.

9.2.4 Disadvantage and Criticism of Triangulation

Triangulation does not come without critique. The disadvantages of triangulation include: (1) the increased amount of time needed in comparison to single strategies, (2) difficulty of dealing with the vast amount of data, (3) potential disharmony based on investigator biases, (4) conflicts because of theoretical frameworks, and (5) lack of understanding about why triangulation strategies were used.

However, probably the largest point of discussion is *philosophical reconciliation,* which is the question of whether different ontological perspectives can be reconciled in the mind of the researcher. This boils down to an ontological question and, therefore, cannot be judged as right or wrong, doable or not doable. It is in the eyes of the beholder whether a researcher can accept that different "versions of the truth" can be reconciled to a larger picture based on integrated worldviews or will remain as separate entities based on different worldviews.

The next section describes a research study that benefited from using a three-point philosophical triangulation to identify new and interesting phenomena that may otherwise have gone unnoticed or considered as methodology-specific phenomena.

9.3 Applying Philosophical Triangulation

9.3.1 Background of the Three Related Studies

The purpose of the three related studies was to understand the relationship between a project methodology and project success in different project governance contexts (Joslin 2015). The literature on project methodologies and project success was based on only a single philosophical perspective. The literature is divided as to whether methodologies that were standardized, customized, or a combination of both led to greater project success (Joslin and Müller 2015b). None of the literature covered the potential moderating effect of different project governance contexts on the relationship between project methodology and project success.

In light of the fact that (1) there was no consistent view of whether methodologies that are standardized, customized, or a combination of both led to greater project success; and (2) combined with the interest in academic research to find alternative research methods that might reduce areas of academic discord (such as the topic on methodologies and project success), a conceptual study was undertaken. Its aim was to understand whether a comparative could be developed and provide an alternative perspective on project methodologies and project success. This comparative was developed under a positivist paradigm that looked for facts and causes using a deductive approach. This comparative showed how a project methodology could be seen in a different perspective. Using the comparative directly resulted in uncovering a number of interesting phenomena described in Joslin and Müller (2015a). The comparative also provided an alternative philosophical perspective in the multiparadigm study.

The second study used a *critical realism* paradigm, applying qualitative methods with the aim of validating the constructs of a theoretically derived research model while gaining insights to steer the direction of a greater study on methodologies, their elements, and their impact on project success. The qualitative study also investigated whether different project environments, notably project governance, impacted the relationship between methodologies and project success. The critical realism paradigm was appropriate, in that it emphasized the need to critically evaluate objects for the purpose of understanding social phenomena (Sayer 1992). Also, critical realism consists of different levels, which addresses the fact that complex social phenomena cannot be explained solely by looking at mechanisms and processes that operate purely on one level (Wikgren 2005).

The third study undertaken used *post-positivism* as the underlying philosophy. Post-positivism assumes that the world is driven mainly by generalizable (natural) laws, but their application and results are often situational dependent. Post-positivist researchers, therefore, identify trends—that is, theories that hold in certain situations but cannot be generalized (Biedenbach and Müller 2011).

9.3.2 Explanation of the Natural-Science to Social-Science Comparative

This section provides a short introduction into the natural- to social-science comparative to aid in the understanding of the findings in the following section. The natural-science comparative model of Joslin and Müller (2013) compares project methodology elements to the genes of an organism. The genes of an organism are the building blocks of the organism (including the observable characteristics), called a *phenotype* (Malcom and Goodship 2001). Genes are switched on and off throughout the life of an organism, which the authors argue is the same concept as the elements of a methodology being applied to a project throughout its life cycle: they are switched on when required, then switched off when not required. The natural-science comparative reifies a project methodology that is considered as the core makeup of a project; therefore, it is responsible for the switching on and off of methodology elements. The project manager is considered to be an environmental variable. Some of the genes in an organism are highly pleiotropic, meaning their impact can be seen in the organism's phenotype—for example, hair color, eyes, and height (Stearns 2010). The comparative explains that the same is true for elements of the applied project methodology. The highly pleiotropic methodology elements noticeably impact the characteristics of project success. In summary, the elements of a methodology and their attributes are compared to and mapped against the attributes of the genes of an organism, and the attributes of a project outcome (product or service) are compared to and mapped against the attributes of a physical organism (phenotype). A detailed explanation, definitions, and mapping tables are described in Joslin and Müller (2015a).

9.3.3 Findings from the Three-Point Philosophical Triangulation

Each of the three studies identified phenomena within and across one or more of the three philosophical perspectives. The phenomena that were identified within only one of the three philosophical perspectives were considered to be methodology dependent. However, phenomena that were identified across two or three philosophical perspectives were considered to be triangulated.

Nine phenomena were identified in total across three philosophical perspectives, and eight of the nine phenomena were triangulated either by the second or third philosophical perspective. Three of the nine phenomena are described in detail below to explain the triangulation of results, and the rest are listed in Appendix A to this chapter (beginning on page 188).

Observed Phenomena Identified at Intersect G (see Figure 9.2) Included the Following:

A comprehensive set of project methodology elements (wherein methodology elements may include tools, techniques, knowledge areas, and capability profiles) positively impacted project success. This phenomena was observed in both the qualitative and quantitative studies, and by using the natural-science comparative, the same phenomena can be explained. The comparative shows that a comprehensive methodology, and its elements can be mapped to the genes of an organism, which ensure the organism is, in Darwinian terms, "fit" (i.e., adapted to the environment and reproductively successful) (Darwin 1859). Therefore, having a comprehensive set of methodology elements ensures full applicability and hence support during the project life cycle. This is the first and simplest example showing that triangulation identifies the different facets of the phenomenon under study.

Observed Phenomena Identified at Intersect E (see Figure 9.2) Included the Following:

The impact of supplementing missing methodology elements to achieve project success that is moderated by project governance was an observed phenomena in the qualitative study. It can also be explained using the natural-science comparative, in which genes of an organism can not only be switched on and off, but new genes can be created (albeit rarely in an evolutionary timeframe) in response to environmental changes (Holliday and Pugh 1975). For example, for hundreds of thousands of years, our ancestors used to see only in black and white; then with the creation of new genes, our ancestors evolved to have color vision (Yokoyama et al. 2014). The trigger for the creation of new genes was due to changing environmental conditions wherein plants, trees, and shrubs started to use color to differentiate their fruits. The creation of new genes was a potential trigger for *stepped* or *punctuated* evolution in a species, so it is more the exception than the norm (Milligan 1986). A more recent and rare example of new genes being created in an organism is the discovery in 2013 of Chinese boy who has the ability to see in pitch black (Scutti 2015). Whether this genetic mutation will proliferate or die out will conform to the laws of Darwinian "fitness" (Darwin, 1859).

The quantitative study (Study 3) did not observe the phenomena at Intersect 2. This may be because the respondents of the study did not experience the situation in which a governance paradigm influenced whether or not an incumbent methodology was supplemented by missing methodology elements.

For information, this particular phenomenon was further investigated using exploratory research to understand whether project governance perhaps had a direct impact on the comprehensiveness of project methodology (Joslin and Müller 2015b). The findings showed that, depending on the project governance paradigm (Müller 2009), a shareholder-oriented, as opposed to stakeholder-oriented, organization is more likely to have incomplete methodology(s), and hence, project managers within a shareholder-oriented organization are more likely to supplement the incumbent methodology. Using a single-paradigm approach would likely have missed this phenomenon and/or missed providing an alternative perspective, therefore making it unlikely that it would have been further researched.

Observed Phenomenon Identified at Segment B (see Figure 9.2):

Segment B in Figure 9.2 denotes the natural-science comparative study in which one of the phenomena observed was how the "core makeup of a project" is defined in terms of the comparative. The comparative sees that the core makeup of a project is its "applied methodology," which contains the *what* to build plus also the information on *how* to build it. What follows is how this phenomenon was derived from the comparative and an explanation as to why it was not triangulated by the other two philosophical perspectives.

In the natural sciences, the core makeup of an organism is not the organism itself but the genes that define how the organism will develop—that is, its phenotype (Dawkins 1974). The genes are part of the chromosomes, which in turn are reflected within the DNA of a cell (Dawkins 1974). The development or growth of the organism, which in the comparative is akin to the project outcome, is decentralized, meaning that every cell is programmed to replicate and develop the organism to the collective good of the organism's genes.

There are, however, master genes that control and monitor the progress of the other genes within their domain to collectively orchestrate the development and maintenance of the organism (Pearson, Lemons, and McGinnis 2005). This master gene concept has been compared to local governance in the social science perspective of projects (Joslin and Müller 2015a). Using the comparative and the mapping tables within the comparative produces the phenomenon that describes the core makeup of a project as the project methodology, but where the *what* to build and then *how* to build it are integrated within the project methodology. This phenomenon derived from the comparative can be explained and understood within the context of the comparative.

The observed phenomenon that the core makeup of a project is the applied methodology was discussed as part of the qualitative study, but there was no common agreement. As many of the people interviewed were project managers or in some way heavy influencers of their projects, they invariably felt that they and their teams were the core makeup of project, even though the knowledge of what to build and how to build it was invariably documented in and applied to a structured methodology. The idea that the core makeup of a project is the applied methodology is not unrealistic, but for the participants of the studies, it was too great a conceptual shift; therefore, the phenomenon was not triangulated.

	One-off Projects	Projects to develop Product/Service which will evolve over time	
Evolution	Generic methodology that may or may not tailored to the type* of project	Knowledge of what to build integrated into methodology elements of how to build	 Closest to Nature
Revolution	Generic methodology that is tailored to the type* of project	Proven methodology based on a previous product or service that is tailored to the type* of project	

*Project types–maintenance, development, research which can result in either a one-off **or** a product/service ongoing development

Figure 9.3 Project Methodology Approaches for Evolutionary–Revolutionary Project Outcomes

One of the questions in the qualitative study asked if there was value to integrating the knowledge of what to build and how to build it into an applied methodology. The majority of the interviews saw the value of this for certain types of projects.

Referring to Figure 9.3, the projects that would benefit the most from integrating the how and the what to build are the evolutionary projects that have long project or service lineage. This models nature, and it was derived from the natural-science comparative. Now consider that for evolutionary projects with long durations, project people come and go, but what remains constant is the knowledge of what and how to build future versions of the product and service as well as incorporating lessons learned. This information will be reflected in the evolving project methodology for that particular product or service. This is another example of a new and interesting phenomenon that may not have been identified and discussed if a single-paradigm approach had been adopted.

In summary, although this observed phenomenon within the comparative was not correlated to other philosophical perspectives—mainly because of engrained beliefs of the participants in the study—phenomenon may be observed and therefore triangulated in future studies that focus on longer-term evolutionary projects.

9.3.4 Summary of Findings

The research question that asked if it was possible to use a philosophical triangulation to identify interesting phenomena, as well as to provide alternative perspectives, has been answered in the above study. A research study using philosophical triangulation can provide many alternative perspectives. Even with a two-point triangulation—that is, two philosophical perspectives—the phenomena under observation can be triangulated, and each can be seen from the other phenomenon's perspective, which provides new and interesting insights, especially when the phenomena are related or correlated in some way.

The natural-science comparative is perhaps one of the most thought-provoking philosophical perspectives because of its objectivity. The comparative is flexible enough to allow many topics of observation to be reified and explained under a natural science perspective and with sometimes counterintuitive findings. For example, topics relating to phenomena that have been observed using the comparative include lessons intentionally not learned, selfish projects, methodologies with a bricolage of competing elements, impotent (generic) methodologies, lone projects (irrespective of size) in a portfolio which are at higher risk of being cancelled than related projects, and lessons learned but fighting for management attention—all of which may or may not be explainable using current philosophical perspectives.

9.4 Discussion

Understanding the need to cross-check the findings of research has been around for over five decades, with the publishing of the first paper on triangulation from Campbell and Fiske (1959). The term *triangulation* is in fact a metaphor taken from the military and applied to natural and social science research (Mathison 1988). The initial expectations of triangulation were at the lower of four levels described by Denzin (1970), which looked at addressing validity and bias. However, with the adoption of new research methods and techniques, the need arose to carry out inter- and intra-method triangulations (Campbell and Fiske 1959). One of the challenges in considering triangulation is the extra effort required to design and run a parallel stream of data collection approaches, additional investigators, and methods. Perhaps this is why only 1.5% of all project management research uses mixed methods—that is, triangulation. This could be an indication of the project management researchers' limited time—many may work only part-time on their research and therefore disregard the benefits of triangulation.

When comparing project management research against business and management research, Cameron and Molina-Azorin (2011) estimate that about 14% of the business and management studies use mixed methods. If there is a link between carrying out different levels of triangulation and observing new phenomena, then project management research is in crisis. If one looks outside the project management research area into the world of practitioner projects, then the low project success rates that are frequently published also imply some form of crisis in the project management field.

A study by Turner, Pinto, and Bredillet (2011) showed that the number of conceptual papers and new techniques in the field of project management has dropped by 10–25% between 1997 and 2007 in two of the three main research journals in project management. This not only indicates a decline in the discovery of new phenomena, but also appears with a stagnating variety of research designs (Biedenbach and Müller 2011; Müller and Söderlund 2015).

This sounds depressing, but there is hope in finding new methods based on transformative research (Drouin, Müller, and Sankaran 2013), such as the natural-science comparative, which was one of the three related studies, and also in using philosophical triangulation as a way to triangulate expected phenomena and discover new and interesting phenomena. The extended use of philosophical triangulation described in this chapter requires three or more philosophical perspectives and not just the two that are typically described in the literature. In doing so, this opens the door to uncovering new phenomena.

However, philosophical triangulation does require an understanding at the outset of a research study that additional effort and rigor are required in the research process to ultimately identify new phenomena in conjunction with the expected phenomena (as part of the overall research study). Using this approach, many of the arguments of the critics of philosophical triangulation are no longer relevant, because the purpose is to discover new phenomena and not necessarily to triangulate expected phenomena as in the studies to date. As a worst-case scenario, once the new phenomena have been identified using this approach, the researcher can always fall back to a single-paradigm approach, which the authors believe is rather unlikely.

9.5 Conclusions

We identified one of the main reasons that, from the practitioner's perspective, project management research produces results of questionable relevance or fails to identify the phenomena of practical relevance. The scarcity of accepted research designs within each research philosophy paradigm limits the variance of research approaches, which reduces the opportunity to identify new phenomena. To address this issue, we first performed a literature review of the four types of triangulation to better understand how triangulation was used and the main benefits it provided, which was primarily in terms of increasing validity and reducing bias. An additional benefit of triangulation, notably at the most abstract level, was philosophical triangulation, which was investigated and provided alternative perspectives on expected phenomena.

We then used three related studies to exemplify the approach of philosophical triangulation, wherein the results of two sets of empirical data (qualitative and quantitative) plus a conceptual study were discussed in different philosophical contexts. The findings show that not only it is possible to create a philosophical triangulation on expected phenomena, but if three or more philosophical perspectives are used, then new phenomena that were not necessarily part of the research hypothesis can be uncovered. This approach to triangulation should provide for richer and more holistic theories. These in turn should help to address the concerns of practitioners by applying the theories that are based on a more integrated view of the project environment.

We can now answer the research question: Philosophical triangulation using three or more perspectives provides for the identification and better understanding of phenomena. Implications for researchers include more detailed understanding of phenomena as a result of better understanding of the different facets of phenomena, theorized from a multitude of ontological perspectives. Theoretical implications include the multilevel triangulation, which allows for better and more realistic theories about phenomena.

This study has, of course, some strengths and weaknesses. The strengths are in the use of existing techniques, which, when combined in a new way, allows for new perspectives toward phenomena. The weaknesses are the limited testing of the application of the new approach. More studies are required that use this approach to identify its benefits and the need for further development. Future research should investigate the use of this new approach for a variety of combinations of philosophies and their triangulations using multimethod and mixed-method designs.

This chapter provided the description and argument for using a new technique in project management research. It is now up to the researchers to use it and reap the benefits from it.

Appendix 9A

Observed Phenomena Across the Three Theoretical Perspectives (Triangulation)

The table below lists the observed phenomena from the related studies described in the paper and which of the phenomena are triangulated to which theoretical perspective.

Table 9A.1 Observed Phenomena Across the Three Theoretical Perspectives (Triangulation)

Observed Phenomena	Intersect	Comments
A comprehensive set of project methodology elements (wherein methodology elements can include tools, techniques, knowledge areas, and capability profiles), all of which positively impact project success.	G	See description within the study.
The impact of supplementing missing methodology elements to achieve project success, which is moderated by project governance.	E	See description within the study.
The core makeup of a project is the project methodology wherein the *what* to build and *how* to build it are integrated within the methodology.	B	See description within the study.

(continues on next page)

Table 9A.1 Observed Phenomena Across the Three Theoretical Perspectives (Triangulation) (cont.)

Observed Phenomena	Intersect	Comments
Necessary and unnecessary complexity.	G	A methodology should contain the necessary complexity to fully support a project during the project life cycle. In the natural-science comparative, there is no such thing as unnecessary complexity; however, in the social science world, a symptom of unnecessary complexity is an ill-fitting methodology that is generic or one that tries to be a one-size-fits-all methodology invariably includes unnecessary complexity. This phenomena was observed in all three theoretical perspectives.
Methodologies with a bricolage of competing elements.	G	A methodology contains a number of elements that can be tools, techniques, processes, knowledge areas, and capacity profiles. For each element selected on a project, there are several other similar elements that are not selected. Therefore, a methodology consists of a bricolage of competing elements; however, once the elements are selected, they work and fit together. This is a similar concept to genes in which the dominant gene is one that has been selected, and on the DNC loci where the recessive genes are the genes that were not successful in being selected (Mendel 1866). The concept of methodologies that comprise a bricolage of competing elements was observed in all three studies.
The impact of a comprehensive set of methodology elements on project success is moderated by project governance.	E	In the natural-science comparative, the equivalent to a comprehensive set of elements is a genome of an organism that allows the organism to be "fit" in the Darwinian sense (Darwin 1859). The environment can impact the genes; and, depending on whether the impacted genes are highly pleiotropic or not, the effect of the environment may be seen in the phenotype—that is, the organism. The qualitative study also identified the impact of governance on whether the methodology elements were considered comprehensive, resulting in an impact on project success. This phenomena was not observed in the quantitative study.

(continues on next page)

Table 9A.1 Observed Phenomena Across the Three Theoretical Perspectives (Triangulation) (cont.)

Observed Phenomena	Intersect	Comments
H2.3. The impact of the application of relevant methodology elements on project success is moderated by project governance.	G–E	The natural-science comparative explains that genes (equivalent in the comparative to elements of a methodology) that are considered relevant and already selected are the dominant genes. These dominant genes can be impacted by the environment. There is little difference in the comparative in explaining the difference between a comprehensive set of genes (elements) and the application of relevant genes (elements). One explanation is that a comprehensive set of genes is the gene pool, and at the point of conception, the bricolage of competing genes are selected. These become the relevant genes, which are then applied (switched on) or not applied (switched off). The phenomenon of project governance impacting the relationship between the application of relevant methodology elements and project success was fully observed in the qualitative study; however, in the quantitative study the phenomena was only partly observed and questionable due to governance being a quasimoderator—that is, a indeterminable effect. Hence G or E.
H1.2. There is a positive relationship between supplementing missing methodology elements and project success.	D	The triangulation only occurs in qualitative and quantitative studies, because there is no natural science equivalent of supplementing missing genes (elements). Organisms conceived with missing genes would typically be terminated soon after conception (Wright 1932). In the social science world of project management, corrections to missing things are allowed and may be encouraged in certain environments. This phenomenon was observed in both qualitative and quantitative studies.
H1.3. There is a positive relationship between applying relevant methodology elements and project success.	G	This phenomenon was observed in all three studies.

Chapter 10

Analysis and Theory Building

This chapter looks at the constructs and hypothesis testing and then discusses the findings of the research by connecting the prestudy with the main study to provide an overarching analysis and discussion. The chapter closes with a theory-building section.

10.1 Construct of Project Success

The topic of success often involves two aspects: success *criteria* and success *factors*. In this study, the focus is on success criteria. Over the past 40 years, there have been many papers written on the topic of how to measure success; however, there is still not consensus among the researchers. Some of the more prominent researchers on success criteria in the project management field include Pinto and Slevin (1988), Morris and Hough (1987), Atkinson (1999), Cook-Davies (2002), Jugdev and Müller (2005), Shenhar and Dvir (2007), and Turner (2008). Their definitions of project success have evolved from the iron triangle in a short-term efficiency perspective to more of a longer-term view incorporating strategic goals of effectiveness and repeatability (Judgev and Müller 2005).

Khan, Turner, and Maqsood (2013) conducted a literature review of project success criteria that spanned 40 years of research and created a scale based on five dimensions:

- Project efficiency
- Organizational benefits
- Project impact
- Stakeholder satisfaction
- Future potential

Khan, Turner, and Maqsood's (2013) model was selected for this study because it is based on the latest literature, which is a superset of the success criteria from the leading researchers on project success. Their model offers a balance between hard and soft factors and measures, as well as presenting both a short- and long-term perspective. The five success dimensions comprise 25 success criteria variables.

Table 10.1 shows the project success variables for each success dimension and provides the results validation.

By using factor analysis, the dimensions in Table 10.1 were reduced to one dimension, called *project success.*

10.2 Construct of PMM Elements

The construct used PMM elements in the quantitative study (Study 3) with the following three dimensions (see also Table 10.2):

1. **MF01.** Comprehensive set of methodology elements
2. **MF02.** Supplemented missing methodology elements
3. **MF03.** Applied relevant methodology elements

The first dimension, a comprehensive set of PMM elements, represents a comprehensive PMM and can be applied to projects so that the PMM does not need to be supplemented. The difference between a PMM and a comprehensive PMM is whether or not the PMM needs to be supplemented by the project manager. Comprehensive methodologies do not need to be supplemented.

The second dimension, supplemented missing PMM elements, refers to an organization's PMM that is not comprehensive and needs to be supplemented with missing PMM elements to achieve a successful project outcome. These elements can be processes, tools, techniques, capability profiles, and knowledge areas.

The third dimension, applied relevant PMM elements (which can include processes, tools, techniques, capability profiles, and knowledge areas), determines whether the relevant PMM elements were applied to achieve a successful project outcome (irrespective of whether or not the PMM was supplemented).

Table 10.1 Project Success Dimensions

Success	Items Included	Results Validation
Project Efficiency	Finished on time Finished on budget Minimum number of agreed scope changes Activities carried out as scheduled Met planned quality standard Complied with environmental regulations Met safety standards Cost effectiveness of work	Shenhar et al. (1997); Cooke-Davies (2002); White and Fortune (2002); Bryde (2005)
Organizational benefits	Learned from project Adhered to defined procedures End product used as planned The product satisfies the needs of users New understanding knowledge gained	Westerveld (2003); Shenhar et al. (1997)
Project impact	Project's impact on beneficiaries are visible Project achieved it purpose End-user satisfaction Project has a good reputation	Westerveld (2003); Shenhar et al. (1997)
Future potential	Enabling of other project work in the future Motivated for future projects Improvement in organizational capability Resources mobilized and used as planned	Cooke-Davies (2002); White and Fortune (2002); Bryde (2005)
Stakeholder satisfaction	Sponsor satisfaction Steering group satisfaction Met client's requirements Met organizational objectives	Westerveld (2003); Shenhar et al. (1997)

The three PMM dimensions described in Table 10.2 include the success factor variables comprised by the PMM dimension. The dimension *comprehensive set of PMM elements* contains 19.8% of the explainable variances for the five PMM success factor variables. The dimension *supplemented missing PMM elements* contains 18.1% of the explainable variances for the five success factor variables. Collectively, the three PMM dimensions explain 55.3% of all of the success factor variables.

Table 10.2 Dimensions of a PMM

PMM	Success Factor Variables	Variance Explained	Accumulated Variance Explained
Comprehensive set of methodology elements	METH01: Comprehensive set processes	19.8%	19.8%
	METH05: Comprehensive set of tools		
	METH09: Comprehensive set of techniques		
	METH13: Comprehensive set capability profiles		
	METH17: Comprehensive set of knowledge areas		
Supplemented missing methodology elements	METH02: Supplemented missing processes	18.1%	37.9%
	METH06: Supplemented missing tools		
	METH10: Supplemented missing techniques		
	METH14: Supplemented missing capability profiles		
	METH18: Supplemented missing knowledge areas		
Applied relevant methodology elements	METH03: Applied relevant processes	17.3%	55.3%
	METH07: Applied relevant tools		
	METH11: Applied relevant techniques		
	METH15: Applied relevant capability profiles		
	METH19: Applied relevant knowledge areas		

10.3 Construct of Project Governance

In the quantitative research Study 3, the construct of project governance is described by two dimensions: shareholder–stakeholder orientation and behavior–outcome orientation. These two axes are the basis of four governance paradigms from Müller (2009). Referring to Table 10.3, the GOVControl factor contains 28% of the explainable variances in the original eight governance questions, and GOVorientation contains 25%. Together they contain 53% of the variance of the 10 governance questions, which is less than the 58.2% that Müller and Lecoeuvre (2014) described for the operationalization of the governance categories of projects.

The "items included" column in Table 10.3 highlights the subjects of the project governance question.

10.4 Hypothesis Testing

10.4.1 Research Model 3 in Study 3

The Research Model 3 in Study 3 has two main hypotheses, each with three subhypotheses.

Hypothesis 1. There a positive relationship between a PMM and project success.

- **H1.1**. There is a positive relationship between a comprehensive set of PMM elements and project success.
- **H1.2**. There is a positive relationship between supplementing missing PMM elements and project success.
- **H1.3**. There is a positive relationship between applying relevant PMM elements and project success.

The hypotheses were tested by using exploratory factor analysis using principle component analysis on the PMM, governance, and success variables to identify the underlying structures and reduce the number of variables to a manageable size while retaining as much of the original information as possible (Field 2009). Validity was tested through unrotated factor analysis for each dimension, which also served as the Haman test to exclude common method bias–related issues, as suggested by Podsakoff and Organ (1986). The results for each of the three concepts gave a Kaiser-Meyer-Olkin (KMO) sampling adequacy value of 0.8 or higher ($p < 0.001$), indicating the data's appropriateness for this analysis.

A control variable was used to filter out spurious effects. It also helped to ensure internal validity. "Years of project experience" was selected as the control variable, because it helped to reduce the confounding effect of variations in a third variable that could also affect the value of the dependent variable. This control variable should also be reflective of experience in using methodologies.

Table 10.3 Construct of Governance

Governance	Items Included	Results Validation	Variance Explained	Accumulated Variance Explained
Behavior–outcome GOVControl	Personnel and procedures Formal and informal PMO Compliance—experience	Müller (2009), Müller and Lecoeuvre (2014)	28%	28%
Shareholder–stakeholder GOVorientation	Decisions Remuneration Profitability and ethics Stakeholder and financial objectives	Müller (2009), Müller and Lecoeuvre (2014)	25%	53%

The results showed that the control variable had no significant effect on the dependent variable, project success; and the PMM factors MF01, MF02, and MF03 were significant ($p \le 0.005$), with an R^2 of 22.3%.

Hypothesis 1, including H1.1, H1.2, and H1.3, is supported when the application of a comprehensive PMM accounts for 22.3% of the variation in project success.

The results support the findings of White and Fortune (2002) and Shenhar et al. (2002), showing that the experience of using a PMM, including the correct choice of tools, techniques, and processes, are two success factors for project success.

Hypothesis 2. The relationship between the project PMM and project success is moderated by project governance.

- **H2.1.** The impact of a comprehensive set of PMM elements on project success is moderated by project governance.
- **H2.2.** The impact of supplementing missing PMM elements on project success is moderated by project governance.
- **H2.3.** The impact of the application of relevant PMM elements on project success is moderated by project governance.

A study by Joslin and Müller (2015b) showed that governance is seen as a major influence on the effectiveness of using PMMs to achieve project success.

For the moderating effect of governance, the findings showed that one of the two moderator factors, GOVorientation, which is the shareholder versus stakeholder continuum, acted as a quasi-moderator. This means that GOVorientation has both (1) an indeterminate relationship between applied PMM elements (MF03) and project success, and (2) the ability to directly influence project success (Sharma et al. 1981). The other two independent variables, comprehensive set of PMM elements (MF01) and supplemented PMM elements (MF02), were not moderated by either of the two moderator factors. Applying the relevant PMM elements' impact on project success is contingent in GOVorientation (shareholder–stakeholder continuum), but it is unclear whether the impact is more on the dependent variable or more on the relationship of an independent-to-dependent variable—thus an indeterminate relationship.

Therefore H2.1, a comprehensive set of PMM elements that is moderated by governance, is *not* supported. Also, for H2.2, supplementing missing PMM elements moderated by governance is *not* supported; and H2.3, the application of relevant PMM elements moderated by governance, is *partly* supported. A table of the research questions, hypotheses, and results are shown in Table 10.4.

Table 10.4 Results of the Hypothesis Testing

Overall Research Question	Main Research Hypothesis	Research Subhypotheses	Research Results	Hypotheses Support
What is the nature of the relationship between a PMM and project success, and is this relationship influenced by project governance?	H1: There a positive relationship between a PMM and project success.	H1.1: There is a positive relationship between a comprehensive set of PMM elements and project success.	A comprehensive set of PMM elements, which includes tools, techniques, processes, knowledge areas, and capability profiles, is positively correlated with project success	Supported
		H1.2: There is a positive relationship between supplementing missing PMM elements and project success	Supplementing missing PMM elements, such as tools, techniques, processes, knowledge areas, and capability profiles, is positively correlated with project success	Supported
		H1.3: There is a positive relationship between applying relevant PMM elements and project success	Applying the relevant PMM elements, which can include tools, techniques, processes, knowledge areas, and capability profiles, is positively correlated with project success	Supported
	H2: The relationship between the project PMM and project success is moderated by project governance.	H2.1: The impact of a comprehensive set of PMM elements on project success is moderated by project governance.	Governance does not moderate or impact the relationship between a comprehensive set of PMM elements and project success	Not supported

(continues on next page)

Table 10.4 Results of the Hypothesis Testing (cont.)

Overall Research Question	Main Research Hypothesis	Research Subhypotheses	Research Results	Hypotheses Support
		H2.2: The impact of supplementing missing PMM elements on project success is moderated by project governance.	Governance does not moderate or impact the relationship between supplementing missing PMM elements and project success	Not supported
		H2.3: The impact of application of relevant PMM elements on project success is moderated by project governance.	Governance quasi-moderates the application of relevant PMM elements to any given project	Partly supported
Does project governance have a positive impact on project success?	H3: There is a positive relationship between project governance and project success	H3.1 There is a positive relationship between the project governance orientation (shareholder–stakeholder) and project success.	Stakeholder-oriented project governance is positively correlated to all five dimensions of project success	Supported
		H3.2: There is a positive relationship between project governance control (behavior–outcome) and project success	Behavior–outcome-oriented project governance does not impact project success	Not supported

Additional findings suggest that project success is more correlated to stakeholder-oriented organizations than to shareholder-oriented organizations. Project success is also associated with organizations that have comprehensive PMMs versus organizations with incomplete PMMs. However, the findings showed that more experienced project managers are needed to effectively apply comprehensive PMMs.

10.4.2 Research Model 4 in Study 4

Research Model 4 in Study 4 has one main hypothesis with two subhypotheses.

Hypothesis 3. There is a correlation between project governance and project success.

- **H3.1.** There is a positive relationship between the governance orientation (shareholder–stakeholder) and project success.
- **H3.2.** There is a positive relationship between governance control (behavior–outcome) and project success.

Using the results of the exploratory factor analysis and validity and reliability testing from the first data analysis for Research Model 1, the two project governance factors, GOVorientation (shareholder–stakeholder) and GOVControl (behavior–outcome), were linearly regressed against project success.

The findings showed that only the GOVorientation (shareholder–stakeholder) factor was significantly correlated to project success, wherein the Beta (0.250, $p <$ 0.001) showed that stakeholder-oriented governance with an R^2 (0.063) accounts for 6.3% of the variation in project success ($p < 0.000$). The results from linearly regressing GOVorientation (shareholder–stakeholder) against each of the five dimensions of project success are shown in Figure 10.1 (page 201).

The success dimension "future potential" is the highest dimension correlated with project governance (adjusted $R^2 = 0.063$; Beta 0.258****). An interpretation of this is that shareholder-oriented governance, through project success, improves organizational capability by fully utilizing its resources and, by successful completion, enables future projects to be selected, resourced, and completed successfully. This drives motivation for future projects, hence improving the organizational capability, which improves the future potential of the organization.

The lowest correlated success dimension of project governance is stakeholder satisfaction (adjusted $R^2 = 0.022$; Beta 0.162**). The explanation for this is that not all stakeholders will personally benefit from the projects, nor will all of the stakeholders approve of the way projects are run, which is in part impacted by the governance approach adopted and the acceptance of the governance by the culture (governmentality) of the organization.

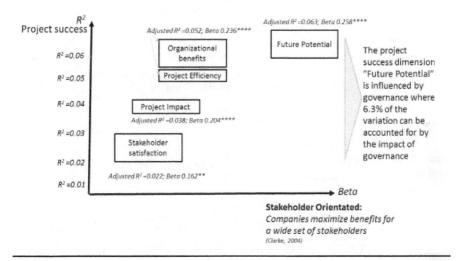

Figure 10.1   GOVorientation (Shareholder–Stakeholder) Factor Impact on the Five Dimensions of Project Success

It is interesting that the GOVControl (behavior–outcome) factor is not significantly correlated, which shows that governance-oriented controls on project processes do not lead to better project success rates.

The summary of research findings is shown in Table 10.4.

10.5 Overarching Analysis and Discussion

This section compares findings from the analysis with the literature and also includes the natural-science comparative to provide a different perspective or a new insight. Figure 10.2 shows a compilation of the key topics and results discussed in the individual papers for this overarching analysis and discussion.

Note: the term *methodology*, when used in the context of projects, has been abbreviated to PMM, meaning project management methodology, and has the same meaning.

10.5.1 Dynamic Set of PMM Elements

PMMs should be seen as a dynamic set of elements consisting of processes, tools, techniques, methods, knowledge areas, and capability profiles that have been

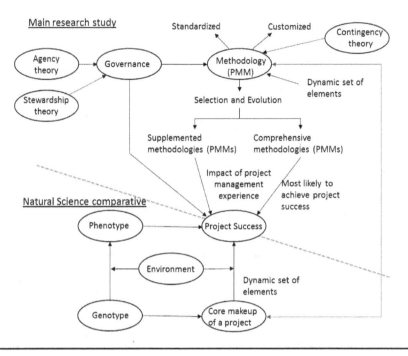

Figure 10.2   Key Topics Raised from This Study

specifically selected for an organization's needs. New elements may be added and old ones replaced according to the needs and demands of the organization's environment and the projects it undertakes. Harrington et al. (2012) refers to a PMM as a "heterogeneous collection of practices that will vary from organization to organization." Toyota has a process that takes individual best practices and tools and, over decades through incremental improvements, adapts them into the product development environment while replacing obsolete or superseded practices and tools (Durward II, Jeffrey, and Allen 1998). The literature does not discuss the dynamic nature of the elements of a PMM, which implies that once a PMM is standardized or customized, it remains that way.

10.5.2 PMM-Related Success Factors

One of the success factors associated with methodologies is the "experience of using PMMs" (Hyväri 2006). This description is probably not precise enough, and the reason for this, as well as a suggestion to revise the success factor description, is given below.

The qualitative study shows that several environmental factors impact the effectiveness of a PMM and its elements. Each PMM element is potentially impacted by the environment; when this interaction and effect exists, it should be under-stood and acted upon by the project manager. These findings are supported in the literature, which advocates the importance of customizing the PMM to the proj-ect environment (Fitzgerald, Russo, and Stolterman 2002; Shenhar et al. 2002).

The environmental factors impacting the effectiveness of a PMM in support-ing project success include project governance, a sponsor's understanding of the need for a PMM, political senior decision makers, culture, and pressure to reduce costs. The quantitative studies did not differentiate between the elements of the PMM—all were grouped and positively correlated with project success. The findings showed that applying the relevant PMM elements is linked to project success. This implies that organizations that have incomplete PMMs are successfully supplementing them with the missing PMM elements (by the project manager) to achieve the desired project outcome—project success. This was also supported in the quantitative findings; however, supplementing PMMs has a lower correlation to project success than do PMMs that are considered to be comprehensive.

One of the project governance factors (GOVorientation) acts as a quasi-moderator, which means that the effect on the relationship between PMM and success was not determinable. The other governance factor (GOVControl) has no moderation effect, but it does impact directly the evolution of a PMM, with consequences resulting in whether it is a comprehensive or incomplete PMM. The effectiveness of the PMM will vary depending on the organization's unique set of influencing factors. Perhaps the success factor termed "the experience of using PMMs" should therefore be reworded to "the experience of applying effective PMMs," because a project manager who uses a PMM that does not take context into account will likely not be effective.

The second PMM-related success factor is the "correct choice of project management PMM/tool" (Fortune and White 2006; Hyväri 2006; Khang and Moe 2008). This is an interesting success factor, because it implies that there is at least one methodology/tool within an organization. What is unclear from the success factor description is whether the organization has made the correct choice for the methodology/tool or whether it is up to the project manager to decide, depending on the type of project. The findings from the quantitative and qualitative studies show that environmental factors, such as governance and culture, influence some of the attributes of a PMM—for example, completeness. In the qualitative study, one of the interviewees stated that "their PMM, which was specifically developed in-house, takes into account the company culture."

PMMs are contingent on context factors, two of which are governance and culture. Selecting the correct PMM does not help achieve project success unless the project manager also has the experience of applying a PMM. Both PMM success factors need to be present to achieve project success; therefore, these factors should be integrated as one success factor and reworded as follows: "experience in selecting and applying effective PMMs."

10.5.3 Core Makeup of a Project

The term *core makeup of a project* was created in the natural-science comparative to mean the applied PMM. This term is not typically used in the project management field, but it would be interesting to see how project managers would view a PMM if they realized that the core makeup of their projects is the PMM (how they are going to build something) and the requirements (what they are going to build). Most project managers probably see their position as the core of a project. In the comparative analysis study, the core is the "what to build and how to build it," wherein a project manager is just one resource among many. Changing the perspective shows that the project manager and all of the other projects resources are enablers, albeit important ones, and nothing more.

10.5.4 Comprehensive PMMs Lead to Greater Project Success Rates

The first quantitative study showed that organizations with comprehensive PMMs have higher project success rates than organizations that need to supplement their PMMs during the project life cycle. The results also showed that it takes more experienced project managers to get the best out of the comprehensive PMMs in order to achieve project success. In the natural-science comparative, the progenotype (like a genotype) evolved to have the perfect combination of elements required to create the organism/project outcome. There is no such thing in nature as supplementing missing genes in a genotype. However, what can happen in nature is the alteration or mutation of a genotype, which means altering the genes. This normally ends in the death of the organism. If the progenotype elements are changed, there is a high risk that the changed elements are suboptimal when compared to the existing elements.

Taking this comparative to the project management world, project managers in an organization with a comprehensive PMM could risk increased project failure when they change any of the proven PMM elements. An organization using a PMM that needs to be supplemented is taking a risk by relying on the project managers to do the right thing in knowing which elements, and with what content, to supplement. This is reflected in the findings of the quantitative study, which showed that supplementing a PMM gives a lower project success rate than using a comprehensive PMM in the first place. These findings are supported by the study from White and Fortune (2002) on current practices in project management, in which they reported that 24% of the respondents found frequent limitations of their project management PMM, and 14% of these reported "unexpected side effects" owing to unawareness of the environment.

10.5.5 The Meaning of Standardized or Customized Methodologies

The literature describes the benefits and the downside to PMMs that are standardized or customized but does not provide information on the origins of the PMM. This is important because only with this information can the terms *standardized* and *customized* be fully understood. For example, for an organization using a standard PMM: Is it based on a generic PMM and called standardized; *or* is it based on a generic PMM that was customized but known internally as a standardized PMM; *or* was the PMM originally developed in-house but known internally as a standardized PMM? It is only with this additional information that the user can understand the true meaning of the terms *standardized* and *customized*.

Another area of confusion about whether a PMM is standardized or customized depends upon the person you ask and where that person is located in the organization. One of the interviewees in the qualitative study who worked at a large USA telecom company talked about a standardized PMM within his/her division, but when asked whether the organization had one standardized PMM, the response what that each division used their own customized PMM, which was considered standardized for the division.

10.5.6 PMM's Influence on Project Success

The correct application of a comprehensive PMM accounts for 22.3% of the variation of project success. This result, from the quantitative findings in Study 3, shows the importance of whether a PMM and its elements are being applied in an effective way. In the qualitative study, one interviewee stated that the organization's PMM was specifically developed in-house and took the company's culture into account, ensuring that the PMM was aligned to the company culture. Another interviewee discussed how the organization had over 40 methodologies that were available to be selected depending on the type of software application, industry, and sector. In section 10.5.2, the two literature-derived success factors associated with project methodologies were refined through discussion to "experience in selecting and applying effective PMMs." A PMM is an organizational asset, but the value of the asset is dependent upon many factors. The maximum value of the asset (PMM) is when the PMM supports the project managers of the organization to make the right decisions in a timely manner, using an efficient and effective combination of processes, tools, techniques, methods, capability profiles, and knowledge areas that give the highest likelihood of project success. Anything less than this reduces the effectiveness of the PMM in supporting project success.

In the natural-science comparative, an organism is considered "fit" because the applicability of the genes' expression of the organism has the best chances for reproduction compared to others in the same evolving species in the given environment (Wright 1932). Every organism in its natural habitat is fine-tuned (adapted) to that environment (Dawkins 1974). Organisms that are low on the fitness landscape invariably become extinct, mainly as a result of their inability to adapt to a changing environment. Examples include the dodo, the Tasmanian tiger, and the dinosaurs.

PMMs serve one purpose—to increase the chances of project success; when they are low on their fitness landscape, for whatever reason, they will eventually be replaced in part or in whole with another competing PMM. However, if the reasons that the PMMs are not achieving their maximum value are not understood, then the fate of the PMM successor is likely to be the same as its predecessor.

10.5.7 Environmental Factors' Influences on the Relationship Between PMM and Project Success

In the natural-science comparative, a genotype comprises thousands of genes, where some genes are highly pleiotropic, meaning they affect multiple functions or characteristics of an organism (Guillaume and Otto 2012). Some of these pleiotropic effects can also be seen in the phenotype—for example, eye color, hair color, and skin color (Stearns 2010). The progenotype (core makeup of a project) is also influenced by the environment; therefore, the traits of the progenotype should be observable in some way in the project outcome and/or the project management outcome.

In the main study, from the qualitative findings, all of the respondents mentioned at least one environmental factor that influences the relationship between PMM and project success. The top five factors raised were project governance, which was the most often cited, followed by four factors that were equally raised—political senior management decisions, culture, resource constraints, and pressure to reduce costs. In the quantitative study (Study 3), one of two project governance factors (GOVorientation, shareholder–stakeholder) was considered to be a quasi-moderator, meaning that the effect on the relationship between a PMM and project success was undeterminable, but the other project governance dimension (GOVControl, behavior–outcome) had an impact on the scope of the PMM in use by an organization.

The findings from the natural-science comparative suggest that environmental factors greatly influence the relationship between a PMM and project success. This is because on the genotype–phenotype (natural science) side of the comparative, the environment greatly influences the development of an organism. Some of the environmental influences can be seen in the phenotype in which the impacted genes (in the genotype) are highly pleiotropic—meaning that the gene is expressed in the phenotype (Guillaume and Otto 2012). Whereas other environmental influences may not be seen in the phenotype, these influences may still impact the organism, but in more subtle ways, such as the ability to ward off infections (Lewtas et al. 1997). This provides a perspective to answering the research question regarding whether a relationship exists between the PMM's elements and project success, and is it influenced by the project context, such as project governance.

The natural-science comparative suggests that the PMM and its elements directly influence project success; however, some of the PMM elements' impact on the characteristics of project success may not be observable (e.g., low pleiotropic PMM elements), but they are still important in achieving overall project success. The project environment will have an impact on the effectiveness of the PMM and its elements in achieving a successful project outcome.

In the first quantitative study (Study 3), the moderating effect is indeterminable from the regression analysis; however, when survey participants of Study 2 were directly asked the question, "Were the use of the PMM elements (including subelements) influenced by the project governance structure?" 38% responded "some influence."

In the qualitative study findings, several environmental factors were mentioned that impact the relationship between PMM and success. One explanation is that using different research paradigms and research methods can provide different results; however, both the natural-science comparative and the findings of the qualitative study suggest that when governance is a major part of the environment, it has some impact on the relationship between PMM and project success (see Section 4.4.6 and Figure 4.6 [page 43]).

For the quantitative study findings, it may be that the impact of project governance is not determinable or that the impact of project governance depends on some other context variables or variables not assessed herein. An explanation for the quantitative finding in Study 3, using the natural-science comparative, is that the PMM elements that were impacted by project governance exhibit only a low pleiotropic effect; therefore, the impact was not detected in the characteristics of project success.

10.5.8 Project Governance Impact on the Completeness of a PMM

In the first quantitative study (Study 3), the finding showed that GOVorientation (shareholder–stakeholder) factor was a quasi-moderator, which suggests that PMM and project success may be contingent on governance, but the results are indeterminate. GOVControl (behavior–outcome) factor was not a moderator but was possibly an exogenous, predicator, intervening, antecedent, or suppressor variable (Sharma et al. 1981). Both governance factors were then tested to see if they mediated the relationship between PMM and project success, and the results were not significant. The author decided to conduct exploratory research to see if the two governance factors directly impacted the three independent PMM factors (MF01, MF02, and MF03). The finding showed that the environment factor governance GOVControl (behavior–outcome) does influence whether or not the PMM is comprehensive (MF01) and whether or not it is supplemented (MF02). This gives an indication that governance may not moderate the relationship between PMM and success, but it does directly influence the evolution of a PMM.

The importance of understanding the history of a PMM and not just its current status of being standardized or not was indicated in the qualitative study (Joslin and Müller 2016b). None of the literature reviewed on PMMs delves into the history or evolution of the PMM, thereby missing an important point of understanding why a certain type of PMM was selected or developed in-house and what the environmental factors are that continue to influence the PMM's development (or evolution). Understanding the history and the evolutionary path of an organization's PMM and the factors influencing it will give a good indication of what it will evolve into and whether this evolution meets or will meet the needs of the organization's projects. It will also provide insight into the skill and personality profile of the project manager required to get the most out of the organization's PMM.

Looking now from a natural science perspective, understanding evolution has given us a picture of how organisms have evolved and how they have adapted (or not) to the environment (Dawkins 1974). Also, the complexity of evolution in natural sciences is increasing (Adami, Ofria, and Collier 2000). This increasing complexity was understood by Lamarck (1838), who believed that the evolution of organisms was a one-way road, which he called "complexity force," or in French, "Le pouvoir de la vie." The topic of complexity is also one of the most discussed topics in project management, in which the development of products or services is becoming ever more complex (Vidal, Marle, and Bocquet 2011). Perhaps one can hypothesize that it is also likely that PMMs will need to become more robust to manage increasingly complex projects, which happens to be a common complaint by many practitioners of today's PMMs (Fortune et al. 2011; White and Fortune 2002).

Without understanding the history of a PMM and the organizational needs it addresses, ill-informed decision makers risk making PMM-related decisions that will negatively impact the ability of a PMM to support the organization's project needs.

10.5.9 Project Governance Direct Impacts Project Success

The literature on project governance shows the diversity of approaches (Müller, Pemsel, and Shao 2014), covering topics such as the optimization of the management of projects (Too and Weaver 2014); the interrelationship of governance, trust, and ethics in temporary organizations (Müller et al. 2013); risk, uncertainty, and governance in megaprojects (Sanderson 2012); governance in particular sectors such as information technology (Weill and Ross 2004); and the normalization of deviance (Pinto 2014). The literature indirectly implies a link to project success, but none have directly measured the influences of different governance orientations on project success.

Governance is pervasive throughout an organization; therefore, it should have degrees of influence on everything that is developed, used, and maintained within and across organizations. The first quantitative study used governance as the moderator variable, and the findings showed that governance had an indeterminate impact on the relationship between applied PMM elements and project success (i.e., a quasi-moderator) (Sharma et al. 1981). However, when project governance was used as the independent variable, the findings showed a direct impact of certain orientations of project governance for both the selection and evolution of a PMM in terms of comprehensiveness as well as project success.

This research endeavor mainly looked at project governance from a positive perspective. The findings from Study 4 ranged from positive to neutral (no impact), depending on the governance orientation. However, the qualitative study (Study 2) shed light on the potentially negative aspects of implemented governance structures. For example, some of the interviewees described misfitting project governance structures that impacted the ability to follow procedures to obtain resources, finalize requirements, test strategies, and provide quality assurance. The findings did not go so far as to suggest actions to enhance the positive aspects and minimize the negative aspects of project governance.

In summary, the finding showed that project governance directly impacts PMM and project success but has an indeterminate impact as a moderator on the relationship between PMM and project success. Could this also be the case with the extant literature on the indirect impact of project governance on project success?

10.5.10 Necessary and Unnecessary Complexity

This section is motivated by the research findings during the development of the natural-science comparative and, more specifically, how nature deals with complexity. It provides an alternative perspective on complexity and the clarification of necessary and unnecessary complexity.

Complexity is a topic that is often discussed in senior management circles inside and outside of the project environment (Hitt 1998). The topic of complexity is especially pertinent to projects, programs, operational systems, and processes (Boyle, Kumar, and Kumar 2005; Joslin 2013).

Evolution has no boundaries, irrespective of whether it is in the fields of natural or social science. Jean-Baptiste Lamarck believed the evolution of organisms was a one-way road, which, as discussed above, he called "complexity force" or "Le pouvoir de la vie" (Lamarck 1838). In social science, the management and development of products or services within projects and programs are also becoming more complex (Vidal et al. 2011).

Many project influencers talk about reducing complexity in projects and systems. This statement is easy to make without really understanding the complexities and challenges of achieving a successful project outcome. If the complexity discussion were moved to the natural science field to build an organism, it is unlikely that the same conversations on complexity would take place. The concern from the project influencers is really about *unnecessary* complexity and not about complexity itself.

Evolution in both the natural and social sciences is resulting in greater complexity (Adami et al. 2000; McShea 1991) but should not be overly complex—as in a practical application of Occam's razor (Hempel 1965). Therefore, the distinction should be made between necessary and unnecessary complexity, and only when this has been done can an intelligent and productive discussion result. Otherwise, there is a risk of removing *necessary* complexity, resulting in reduced organizational performance. This would be called *devolution* or *backward evolution,* which *Scientific American Magazine* nicely frames thus: "from a biological perspective, there is no such thing as devolution" (Dougherty 1998).

10.5.11 Contingency Theory—Theoretical Lens for Qualitative Study 2 and Quantitative Study 3

The comparative analysis is built around contingency theory wherein the environment uniquely influences the elements of a PMM as well as the product, service, or applicant in its operational life. Contingency theory was also used as a theoretical lens for the main study.

The findings from the interviews in the qualitative study show that project governance was the environmental factor most often mentioned as impacting the effectiveness of the applied PMM. Examples were given of misfitting project governance structures impacting the ability to follow procedures to obtain resources, finalize requirements, test strategies, and perform quality assurance. The interviewees did not go so far as to suggest actions to enhance the positive aspects and minimize the negative aspects of the environmental governance factor, but success was contingent on project governance.

In the quantitative (Study 3), governance was found to have a direct influence on whether or not an organization has a comprehensive PMM, and whether or not the elements are supplemented. The findings also showed that organizations with comprehensive PMMs have higher project success rates than those that do not, so success is contingent on project governance. In addition, project governance was a quasi-moderator, which suggests that PMM and project success may be contingent on project governance, but the results are indeterminate.

Fitzgerald, Russo, and Stolterman (2002) noted that the most successful PMMs are those developed for specific industries or organizations, which suggests that contingency theory was a good choice as the theoretical lens for this study.

10.5.12 Agency Theory, Stewardship Theory— Theoretical Lens for Quantitative Study 4

There are several management theories that have been used in the governance area—notably, agency theory, transaction cost economics, stakeholder theory, shareholder theory, stewardship theory, and resource dependency theory, which have all helped to explain observed phenomena (Yusoff and Alhaji 2012).

Agency theory and stewardship theory were selected as the theoretical lens in the quantitative Study 4. Agency theory describes a relationship between two parties (the principal and the agent), in which both actors are perceived as rational economic actors who act in a self-interested manner (Mitnick 1973). Agency theory is particularly relevant in the field of project management, because there are many principle–agent relationships in a project supply chain. Stewardship theory is often considered the contra of agency theory (Donaldson and Davis 1991), and therefore is also relevant to this study, especially when considering projects that are developed within a stakeholder-orientation governance paradigm.

Project managers (agents) are tasked with complex projects and need to get things done; therefore, flexibility and trust is required from their principle (Turner and Müller 2004). If trust is present, this implies that project managers are acting in a stewardship role on behalf of their principle (manager or project sponsor), which they (manager or project sponsor) should also be acting as a steward on behalf of the stakeholders (including shareholders) of the organization. The findings of the study show that stakeholder governance is significantly correlated to project success, whereas the (behavior–outcome) control-oriented governance has no impact on project success. This is aligned with stewardship theory.

Out of the five success dimensions (project efficiency, organizational benefits, project impact, future potential, and stakeholder satisfaction), the lowest correlated success dimension to governance is stakeholder satisfaction. One explanation for this is that not all of the stakeholders will personally benefit from the projects, nor will all of the stakeholders approve of the way projects are run, which is in part impacted by the governance approach adopted.

The findings could imply that principle–agent issues exist that are impacted by the governance of the project when these agents do not personally benefit during the life of the project and/or through the project outcome. This is especially true when the project goals provide increased transparency and/or processes and controls that reduce the opportunity for personal gains.

From a natural science perspective, can agency theory or stewardship theory be explained by the comparative? Agency theory suggests self-interest only, which fits the Darwinian laws of fitness (Darwin 1859). Dawkins (1974) would also agree with the concept of the selfish gene that altruism does not exist in a natural world.

Several studies, which are based on game theory, also support this, including the prisoners' dilemma, and show why two individuals might not cooperate, even if it appears that it is in their best interests to do so (Nowak and Sigmund 1993). Is it possible then that stewardship theory has a place in natural science? Only if the organisms live in a "social" society, in which *social* is defined in terms of the structure and order of the society (Lin and Michener 1972). If stewardship theory has a place in the natural science world, it would be at the top level of social organisms, called "eusocial," notably, bees, ants, and other colony-oriented organisms (Kramer and Schaible 2013).

In this case, however, the greater good of the colony is for the greater good of the individual organism passing on the collective genes of the colony, which are all derived from the queen of the colony. It is up to the reader to debate whether there is really a place for stewardship theory in the natural world.

Agency theory and stewardship theory have helped us to understand and interpret the findings of the study and, while doing so, created new questions for further research (e.g., is stakeholder satisfaction contingent on a stewardship-orientation environment?).

10.5.13 Theory Building

Theory building occurs in two stages: the *descriptive* stage and the *normative* stage. Within each of these stages, theory builders proceed through three steps (Carlile and Christensen 2005) (see Figure 10.3). Kuhn (1970) observed that during descriptive theory building, confusion and contradiction were typically the norm.

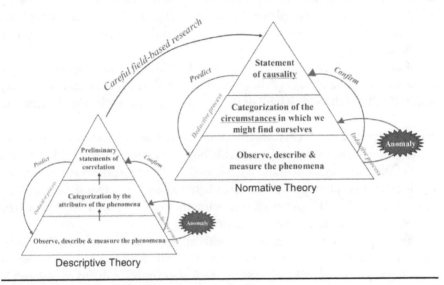

Figure 10.3   The Three Steps in Descriptive Theory and Normative Theory (*Source:* Carlile and Christensen 2005, used with permission)

The natural-science comparative follows the descriptive theory triangle in Figure 10.3, whereas the main study followed the normative theory triangle. Because the natural-science comparative used the same variable as the main study comparisons—that is, *methodology*—observations could be made in one study and then determined as to whether the same phenomena can be seen and explained in the other studies. The comparative analysis starts with a mapping of concepts and terminology and, in doing so, explains why phenomena in genomics (study of genetics) can be compared with practices, behaviors, and established thinking in project management. The natural-science comparative follows a deductive process, working from the top of the pyramid to the bottom. The main study was mostly deductive, except for the qualitative study with its interviews, which were inductively analyzed. For both the qualitative and quantitative studies, the normative pyramid was followed, mainly from the top to the bottom.

These research findings have helped to quantify the impact of a PMM at the collective element level, similar to the genes and the impact on an organism. Contingency theory has been used as the theoretical lens supported in the first three studies, in which environment has a direct and indirect influence on the selection of a PMM, the evolution of a PMM, and the influence on the relationship with project success. The natural-science comparative indicates that the elements of a PMM actually compete against each other to be selected within a PMM. A practical example of this is the choice of similar tools, templates, techniques, and processes that are available, and only the best one of each group will be selected.

This competitive aspect of the elements of a PMM was not part of the original scope in the main part of the research. Instead, the first quantitative study results (Study 3) showed that all of the relevant and applied PMM elements are positively correlated to project success. This is consistent with the natural-science comparative in which, once the elements are selected, they no longer compete—instead, they work together. This notion of competing PMM elements comes from the comparative and is explained by the conception of an organism. During the conception period, the genes fight to be selected (dominant), so once attached to the loci (DNA), from this point onwards, they work together to the collective good of the organism (Mendel 1866).

The comparative analysis also shows that all the elements of the PMM will be impacted from the project environment, but not all of the impact will be seen as traits in project success. This is because each element has a different pleiotropic level, meaning that only the highly pleiotropic elements are likely to be seen as an impact of the characteristics of project success—that is, in one or more project success criteria. PMM elements that have a big impact on the likelihood of project success are called *success factor elements*—for example, realistic schedule, efficient cost control, and accurate budgeting.

The research model for Study 3 has been redrawn (see Figure 10.4) to include the direct interaction of one of the governance factors (GOVControl) on two of the PMM factors (MF01 and MF02). The first factor, MF01, determines whether the organization's PMM is comprehensive in terms of its elements. The second factor, MF02, determines whether the project manager supplemented the organization's PMM knowing that it was not comprehensive and/or felt some of the personal PMM elements were better suited to the project at hand. GOVorientation (shareholder–stakeholder) is a quasi-moderator, meaning it has an indeterminable effect on the moderation of MF03 and project success.

The first quantitative study (Study 3) also conducted exploratory research and showed that the environment factor for governance, GOVControl (behavior–outcome), does influence whether or not the PMM is comprehensive. Even though governance may not moderate the relationship between PMM and

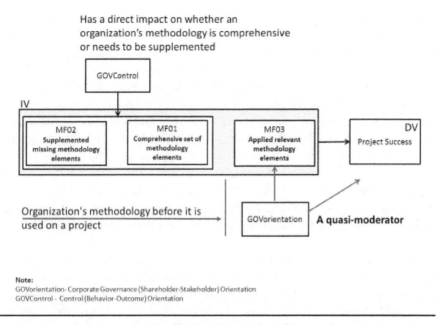

Figure 10.4   Redrawn Research Model Indicating the Influence of Project Governance on the Independent and Dependent Variables

success, it does directly influence the evolution of a PMM in terms of whether or not it is comprehensive—that is, it does or does not need to be supplemented. The qualitative study showed the importance of understanding the history of a PMM in terms of its evolution, regardless of whether a PMM evolved from a generic, standard PMM into varying levels of customization, or a PMM was developed in-house and evolved into varying levels of customization.

The quantitative Study 4, built on the findings (and questions) from quantitative Study 3, was conducted to understand whether different governance orientations directly impact project success. The theoretical lens used was both agency theory and stewardship theory, which are more applicable than contingency theory, which was used in the first three studies. The findings of Study 4 show that stakeholder governance is significantly correlated to project success, whereas the (behavior-outcome) control-oriented governance has no impact on project success. This is aligned with the concept of stewardship theory. Out of the five success dimensions (project efficiency, organizational benefits, project impact, future potential, and stakeholder satisfaction), the lowest correlated success dimension to governance is stakeholder satisfaction. One explanation for this is that not all stakeholders will personally benefit from the projects. The findings could imply that principle–agent issues exist that are impacted by the project and/or project outcome.

As discussed above, from a natural science perspective, agency theory suggests self-interest only, which fits the Darwinian laws of fitness (Darwin 1859). Dawkins (1974) would also agree with his concept of the selfish gene that altruism does not exist in a natural world, supporting the concept of agency theory.

Stewardship theory, on the other hand, was found to have little or no place in natural science. If stewardship theory does apply, it is only if the organisms live in a social society, where "social" refers to the structure and order of the society (Avilés and Purcell 2012). Even then, if stewardship theory were to explain natural science phenomena, it would have to be at the top level of social organisms, which are called *eusocial*—notably, bees, ants, and other colony-oriented organisms (Linksvayer 2010). So, would organisms choose the greater good of the colony over personal gain? Yes, but only because the individual organisms ensure the passing of the collective genes of the colony (which are all derived from the queen of the colony), therefore implying some form of self-interest.

In summary, this chapter detailed the constructs for three variables: project success, PMM, and project governance. The hypotheses were described with supporting literature, and the prestudy and main study were brought together to provide alternative perspectives, challenges, and questions to the observed phenomena coming out of the other three studies.

The next chapter concludes the book, shows the contributions to both the academic and management communities, and provides suggestions for future research.

Chapter 11

Theoretical and Practical Implications

A summary of the research process, overview of the findings, and revelation of whether the objectives of the research were met are followed by answers to the research question. The strengths and limitations of the study are highlighted. The theoretical and practical contributions of the research are listed, concluding with recommendations for future research.

11.1 Summary of the Research Process

The prestudy (see Chapter 5) uses a deductive approach applied under a positivist paradigm wherein the observer looks for facts and causes. For the main part of the research, the qualitative study (see Chapter 6) used critical realism, which assumes that reality is mostly objective, but social constructions are recognized and must be outlined in an objectivist way (Alvesson and Sköldberg 2009). Then the following two quantitative studies (Chapter 7 and Chapter 8) used post-positivism, which "assumes that the world is mainly driven by generalizable (natural) laws, but their application and results are often situational dependent. Post-positivist researchers therefore identify trends—that is, theories which hold in certain situations, but cannot be generalized" (Biedenbach and Müller 2011).

For the prestudy, once the philosophical perspective was clarified, the foundations of the comparative were defined, and a literature review was carried out on the foundation of the comparative (Universal Darwinism), which was then extended to areas in biology—more specifically, genomics—in which the characteristics of a PMM and project outcome were mapped. The results of the comparative analysis were written into a book chapter (Joslin and Müller 2013); then a theory-building section and detailed comparative literature were added, which were presented at the EURAM conference (Joslin and Müller 2014); and subsequently the comparative was published in Joslin and Müller 2015a.

For the main study, the research was executed through three stages using mixed methods. The first stage was a qualitative study (see Chapter 6), in which 19 semistructured interviews were conducted. The interviewees were project, program, and senior IT managers from seven industries across four countries—Switzerland, Germany, UK, and the USA—who all have detailed knowledge of their organization's PMM(s). The aims of the study were to qualitatively validate the constructs of the research model, gain agreement in the use of the term *PMM elements*, and gain additional insights, such as the importance of understanding the PMM source and level of customization. The guidelines of Miles and Hubermann (1994) were followed during the research. The results were written into a research paper that was presented at the PMI Research and Education Conference in Portland, Oregon (Joslin and Müller 2016a), then improved and subsequently published in the *International Journal of Managing Projects in Business* (*IJMPB*) (2016b). The *IJMPB* paper is the basis of Chapter 6.

The second stage of the main study was the quantitative study (see Chapter 7). A web-based questionnaire was developed to collect information on methodologies, project success, and governance paradigms. The scales for project success were taken from Khan, Turner, and Maqsood (2013), and the scales for governance came from Müller and Lecoeuvre (2014). The PMM scales were based on the data collected in the qualitative study, and factor analysis was used to determine the internal consistency of the scales. The process to carry out the data analysis followed the guidelines of Hair et al. (2010). The results were written into a research paper and published in the *International Journal of Project Management* (*IJPM*).

The third stage of the main study was also a quantitative study (see Chapter 8) and utilized the data, scales, and some of the open questions from the first quantitative study. Analysis was done through factor and linear regression analyses. The process to carry out the data analysis followed the guidelines of Hair et al. (2010). The results were written into a research paper and published in *IJPM*.

11.1.1 *Overview of the Research Findings*

Chapter 5 shows how it is possible to create a new comparative by mapping concepts and terminology, and, in doing so, it explains why phenomena in genomics (the study of genetics) can be compared with practices, behaviors, and established thinking in project management. The chapter's literature review includes the importance of comparatives in research and the steps that were taken over the past 30 years to improve them. The research discovers that the structure of biology from a cell level down to the gene is analogous to a library in which the lowest level is a page element that contains a piece of knowledge. The foundation of the comparative is based on Universal Darwinism, which is an extension of Darwin's evolutionary process. The discovery was made that a PMM and its elements could be compared against a genotype and its genes where an applied (lived) PMM is the core makeup of a project. There is also a theory-building section that uses the concept of complex adaptive systems (CAS), which have been well researched over the past 50 years by Holland (1992), Mitchell (1998), and Lansing (2003), among others. The challenge to theory building in this comparative is that the comparative spans the natural and social sciences and functions at both the suborganism level (genes) and the organism level (phenotype). Likely, so does CAS. To help with the explanations, the PMM was reified to what it would be like if the elements of a PMM were competing to be selected for a project. With the comparative built, three scenarios were described and explained using the model—selfish projects, lessons intentionally not learned, and competing PMMs—with the bricolage of individual elements through use and copy across PMMs.

A paper based on the prestudy was presented in June 2014 at EURAM in Valencia, Spain; then it was improved and subsequently published by PMI's *Project Management Journal®* for the January 2015 issue (Joslin and Müller (2015a).

Chapter 6 describes the findings of the qualitative study. The study establishes that there is a positive relationship between the elements of a PMM and project success, and that the effectiveness of the PMM varies according to environmental factors. A number of environmental factors were mentioned, with governance being the most often cited. From the interviews, the importance in understanding the origin of a PMM was evident. Just knowing if the incumbent PMM is standardized or customized is not sufficient, because a PMM could be generic, thus standardized; generic then customized, thus standardized; or customized, thus standardized. These steps also help to understand the evolution of the PMM in the organization and what is meant by standardized or customized. This study was presented at PMI's Research Conference in Portland, Oregon, in July 2014, improved and subsequently published by the *International Journal of Managing Projects in Business* (*IJMPB*) (2016b).

Chapter 7 discusses the finding of the first quantitative study. A new PMM scale was created using three factors, and all three factors were positively corrected to project success. Project governance as a moderator was represented by two factors: shareholder–stakeholder and behavior–outcome. Shareholder–stakeholder was found to be a quasi-moderator on one of the three independent factors: MF03 (applying relevant PMM elements). The second project governance factor, behavior–outcome, was not a moderator but an exogenous, predicator, intervening, antecedent, or suppressor variable (Sharma, Durand, and Gur-Arie 1981). The behavior–outcome project governance factor was regressed directly against the three independent factors, and findings showed that outcome-oriented organizations are more likely to supplement missing PMM elements, as required, than those that are more compliance-oriented and have had a complete PMM from the outset. Project governance may have an indeterminable effect when a PMM is applied, but before a PMM is applied, project governance impacts the selection of the PMM and whether it is comprehensive or needs to be supplemented by the project manager during a project life cycle.

The results were written into a research published by *IJPM* (2015b).

Chapter 8 discusses the finding of the second quantitative study, which utilized the data, scales, and some of the open questions from the first quantitative study. This study explores the impact of project governance on project success. The findings show that a stakeholder-oriented project governance accounts for 6.3% of the variation in project success, and that project governance structures that are more control behavior–outcome oriented have no impact on project success.

The results were written into a research paper published by *IJPM* (2016c).

Chapter 9 discusses that the scarcity of accepted research designs within each research philosophy paradigm limits the variance of research approaches, which reduces the chances to identify real new phenomena. The chapter proposes that researchers use triangulation of alternative research philosophies to identify interesting new phenomena, provide alternative perspectives to complex problems, and gain a richer and more holistic understanding of complex project management problems

The results were written into a research paper published by *IJPM* (2016d).

11.1.2 Hypothesis Testing

There were three main hypotheses and eight subhypotheses in this research endeavor.

For the first hypothesis, H1, *there is a positive relationship between a PMM and project success;* the research showed that 22.3% of the variation of project success is accounted for by the correct application of a PMM with a significance of ($p \leq 0.005$).

Table 11.1 Research Findings Overview

Chapter	Comments	Journal/Conference Proceedings
Chapter 5: New Insights into Project Management Research: A Natural Sciences Comparative	Natural- to social-science comparative including theory-building section and comparative section	EURAM 14th Annual Conference, Valencia, Spain, June 2014 Joslin, R., & Müller, R. (2015a). New insights into project management research: A natural sciences comparative. *Project Management Journal*, 46(2), 73–89.
Chapter 6: The Impact of Project Methodologies on Project Success in Different Project Environments	Qualitative part of the PhD	PMI Research Conference, Portland, Oregon, July 2014 Joslin, R., & Müller, R. (2016b). The impact of project methodologies on project success in different project environments. *International Journal of Managing Projects in Business*, 9(2), 364–388.
Chapter 7: Relationships Between a Project Management Methodology and Success in Different Project Governance Contexts	Quantitative part of the PhD	Joslin, R., & Müller, R. (2015b). Relationships between a project management methodology and project success in different project governance contexts. *IJPM*, 33(6), 1377–1392.
Chapter 8: The Relationship Between Project Governance and Project Success	Quantitative research based on data obtained from the online survey	Joslin, R., & Müller, R. (2016c). The relationship between project governance and project success. *IJPM*, 34(4), 613–626.
Chapter 9: Using Philosophical and Methodological Triangulation to Identify Interesting Phenomena	Qualitative part of the PhD	Joslin, R., & Müller, R. (2016d). Identifying interesting project phenomena using philosophical and methodological triangulation. *IJPM*, 34(6), 1043–1056.

The subhypotheses:

- **H1.1.** *There is a positive relationship between a comprehensive set of PMM elements and project success*—was supported. Additional analysis showed that PMMs that are comprehensive have higher success rates than PMMs that need to be supplemented. This implies that organizations with comprehensive PMMs invest in updating their PMMs through lessons learned and/or new PMM elements that are more appropriate than the ones they replace.

- **H1.2.** *There is a positive relationship between supplementing missing PMM elements and project success (MF02)*—was supported. Incomplete PMMs create a risk that, unless supplemented, will negatively impact project success. Organizations with incomplete PMMs would have to rely on the experience of their project managers to determine how, what, and when to supplement so as to ensure a well-executed and successful project.

- **H1.3.** *There is a positive relationship between applying relevant PMM elements (MF03) and project success*—was supported. Having a comprehensive PMM and supplementing a PMM when PMM elements are missing is important; however, unless the PMM elements are relevant to the particular project and applied in an effective way, the chances of process success are reduced. The Pearson correlation of MF03 to project success shows the highest correlation (0.385) compared to MF01 (0.196) and MF02 (0.168), which confirms its importance as a key project success factor.

The second hypothesis, H2, *the relationship between PMM and project success moderated by project governance,* was partly supported.

The subhypotheses:

- **H2.1.** *The impact of a comprehensive set of PMM elements on project success is moderated by project governance*—was not supported. Having a comprehensive model is of little value until it is applied on a project; therefore, governance would not play a part in impacting the relationship between PMM and project success until the PMM is applied. So the findings are logical. This does raise the question as to whether governance directly impacts the selection and evolution of a PMM, which is answered in the discussion section (see Section 6.5, page 101).

- **H2.2.** *The impact of supplementing missing PMM elements on project success is moderated by project governance*—was not supported. The comments for H2.1 also apply to H2.2.

- **H2.3.** *The impact of application of relevant PMM elements on project success is moderated by project governance*—was partly supported. One of the two moderating factors, GOVorientation (shareholder–stakeholder), was observed to be acting as a quasi-moderator and not a full moderator. The second proposed moderator, GOVControl (behavior–outcome), was not a moderator but possibly an exogenous, predicator, intervening, antecedent, or a suppressor variable (Sharma et al. 1981). Therefore, Hypothesis 2 is only partly supported.

The third hypothesis, H3, *there is a positive relationship between project governance and project success,* was supported.

The subhypotheses:

- **H3.1.** There is a positive relationship between the governance orientation, GOVorientation (shareholder-stakeholder), and project success—was supported. The findings showed that shareholder-oriented governance is positively correlated with project success and its five success dimensions (stakeholder satisfaction, project impact, project efficiency, organizational benefits, and future potential). Project governance that is aligned to the stakeholders who influence and directly support the project is far more likely to be accepted from a governmentality perspective of project governance (Foucault 1979) than project governance that is only aligned to one stakeholder (i.e., the shareholder). Project governance needs to fit in and be accepted by the culture of the organization in order to have the greatest impact; therefore, a multiview perspective is important, which implies a stakeholder-oriented governance structure.

- **H3.2.** *There is a positive relationship between governance control (behavior-outcome) and project success*—was not supported. This subhypothesis was not supported, which challenges the literature on frameworks like the Capability Maturity Model Integration (CMM Integration[gfv]). Maturity models such as CMM Integration or OPM3® are based on the premise that stronger project process controls increase the chance for project success. The findings show that requiring people to follow project processes does not necessarily lead to better project results. Instead, understanding and managing the diverse needs of project stakeholders, which is reflected in a stakeholder-oriented governance structure, leads to the highest chance for project success.

A summary of the hypotheses testing results are shown in Table 10.4 in the previous chapter (page 198).

11.1.3 Answers to Research Questions

There were two core research questions, one relating to the prestudy and the second to the main part of the research. The research questions have already been summarized in the research findings and as a consequence of the research papers. The following are answers to the research questions.

Prestudy

The prestudy (Study 1) research question was formulated as follows:

How can a natural science perspective be used in understanding social science phenomena where methodology is the social science phenomena under observation?

By developing a natural-science comparative to social-science comparative. It is achieved by mapping concepts and terminology; and in doing so, it explains why phenomena in genomics (the study of genetics) can be compared with a PMM and the resulting outcome of a product or service. Examples of project management phenomena derived from the comparative include selfish projects, lone projects with an increased risk of cancellation, competing PMM elements, and lessons intentionally not learned.

Main Study

For the main research, the core research question was formulated as follows:

What is the nature of the relationship between the PMM, including its elements, and project success, and is this relationship influenced by the project environment, notably project governance?

The findings of the qualitative and quantitative studies showed a positive relationship between PMM and its elements and project success. The qualitative study findings also found a link where project context influences the relationship between PMM and success, whereas a number of context factors were given—for example, project governance, senior management politics, culture, and budget cuts. Project governance was the most frequently mentioned context factor.

The second part of the mixed-methods quantitative study (Study 3) refined the research question into:

What is the nature of the relationship between the PMM's elements and project success, and is this relationship influenced by project governance?

PMM elements have a positive relationship with project success, but only one of the two governance factors, GOVorientation (shareholder–stakeholder), showed a quasi-moderating effect, which, according to Sharma et al. (1981), is indeterminable. The second governance factor GOVControl (behavior–outcome) is not a moderator but is possibly an exogenous, predicator, intervening, antecedent, or a suppressor variable.

In summary, the findings from the three studies show that a PMM should be seen not as a homogeneous entity, but as a living and evolving heterogeneous set of elements (processes, tools, techniques, capability profiles, methods, and knowledge areas) that are heavily influenced by the environment. Environmental influences impact not just the appropriateness of the PMM elements for any given project, but also the original PMM selection process in terms of the type of PMM selected and how the PMM evolved in terms of the characteristics of the PMM (comprehensiveness). Organizations that take a holistic view into understanding the factors influencing the PMM selection, PMM evolution, and PMM appropriateness (including the elements) for any given project are likely to see improved project success rates with the added benefit of likely reduced complaints about the inappropriateness of a PMM.

The third part of main study the mixed method quantitative research (Study 4) looks at the impact of project governance on PMM and project success. The following research questions are asked:

1. *What is the relationship between project governance and a PMM?*
2. *What is the relationship between project governance and project success?*

Behavior–outcome-oriented governance has a direct, significant impact on two of the three interdependent PMM factors (MF01 and MF02), which shows (1) that governance does influence, in some way, the selection and evolution of a PMM and (2) whether the PMM is comprehensive or not. Organizations that are more outcome-oriented have incomplete PMMs, whereas behavior-oriented organizations have more comprehensive PMMs.

Stakeholder-oriented project governance structures accounts for 6.3% of the variation in project success, so stakeholder-oriented governance is correlated to project success. However, for organizations having project governance structures that are more behavior–outcome oriented, their project governance structures have no impact on project success.

11.1.4 Theoretical Implications

Contingency theory is applicable to organizational PMM selection, evolution, selection per project, supplementation, and customization according to the project's environment.

Agency theory and stewardship theory help to explain governance-based phenomena directly relating to PMMs and project success.

Governance plays a quasi-moderating effect on the applied PMM and directly impacts the establishment of the PMM and how it evolves to be comprehensive and/or needs to be supplemented before project use.

A PMM's effectiveness is continuously being impacted by the project's environment throughout the project life cycle, which impacts project success.

The natural-science comparative suggests that there may be other types of moderators that influence the relationship between PMM and project success, therefore indicating the need for further research.

Project governance directly impacts project success when the governance approach is more stakeholder oriented, reflecting the importance of governmentality.

Project governance that is more control oriented, irrespective of whether it is behavior or outcome focused, has no impact on project success, therefore challenging project process structures that are more control oriented.

11.1.5 Managerial Implications

The managerial implications address both the project manager and senior managers of organizations.

Organizations that are more focused on shareholders than stakeholders have a lower probability of project success; therefore, a project portfolio manager who knows his/her organization's governance paradigm and the implications on current and future projects may help influence, shift, or create local project governance paradigms that are more conducive to success.

Organizations that are more outcome oriented supplement their PMM more than organizations that are more process/compliance oriented, where the latter organizations are more likely to have a comprehensive PMM than the former organizations. The organizations controlled by outcomes expect the project manager to perform and supplement the PMM as necessary so as to meet the goal of shareholder value. These organizations should only recruit senior project managers who have the experience to determine what to supplement in a PMM in order to achieve project success.

Organizations that have a more comprehensive PMM also need experienced project managers to ensure the achievement of high success rates. By understanding the governance paradigm and the state of the evolution of the organization's PMM, a program or project portfolio manager will have a good indication of the project management skills and, especially, the experience necessary for a successful project outcome.

When project success rates are dropping and lessons learned indicate a misfitting PMM, understanding the governance paradigms and the risks associated with the evolution of PMMs within each governance paradigm may provide valuable information as to the root cause of the problems.

Organizations that enforce a strict governance approach for process compliance reduce the opportunity for maximizing their organization's chances of project success, unlike organizations that implement a stakeholder-oriented project governance, which is correlated to project success.

11.2 Strengths and Limitations

One strength of the study has been to provide insight into the benefits of understanding the origins of an organization's PMM as well as the governance paradigms to obtain an understanding of why the organization's PMM has evolved to its present time. Another strength of the study is to use the natural-science comparative to determine whether the findings and potential explanations can also be supported in the comparative.

To the best of the author's knowledge, this is the first study to quantitatively assess the contribution of PMM usage to project success.

A limitation of this study was the questionnaire distribution method, in which snowball-and-convenience sampling does not allow for questionnaire distribution by industry, project type, or geography. Using professional associations such as IPMA® and PMI for distribution of the questionnaire may also exclude project managers and other applicable respondents who are not part of these professional associations. Other limitations are the inconclusive findings on the role of governance as a moderator or quasi-moderator and the fact that the natural-science comparative has not been tested outside of this study.

Viewing a PMM in terms of elements that may exist within a hierarchy may be a strength or a weakness. The strength of the approach is that it allows comparisons to be made using the natural-science comparative as well as the concept of elements exhibiting individual and collective group effects on the characteristics of project success. Also, by using the term *elements,* a more neutral feeling is allowed from the perspective of the project management, whether elements are kept or replaced with something that is more appropriate for the project at hand. The limitation of this approach is that it requires a project manager or reader to shift his/her view of PMMs from a homogeneous entity to something that is a heterogeneous collection of elements.

11.3 Recommendations for Future Research

11.3.1 Natural-Science Comparative

One recommendation for future research is to apply the comparative model in existing research areas in order to understand how it performs in terms of supporting current findings, challenging current findings, and discovering new findings. It would also be important to understand the limitations and strengths of the comparative model.

Another recommendation is to extend the comparative model along the attribute dimensions to allow a broader scope of applicability. For example, the attribute dimension "collaborate" can be extended to include social organisms (Danforth 2002; Simon 1960), which will provide insights into understanding why independent projects in a project portfolio may be at greater risk of being canceled or put on hold than linked or related projects.

11.3.2 Main Study

The main study showed that it was important to understand the origins of an organization's PMM, how it evolved, and why it evolved in the context of its environment. For future research, this could include understanding of the potential moderating or mediating factors between PMM and project success. These factors could include organizational shocks, such as cost-cutting programs, transformative environments, and politics. Another research area is to understand whether moderators and/or mediators of the relationship between PMM and project success have an impact on the decision to adopt or create a PMM and/or influence how it evolves.

Another angle for research is to focus on one PMM element, such as a process or a tool, and then develop a subelement structure and determine the impact of the governance paradigms on this element and its related subelements.

In summary, the findings from the three studies show that a PMM should be not seen as a homogeneous entity but as a living and evolving heterogeneous set of elements (processes, tools, techniques, capability profiles, methods, and knowledge areas) that are heavily influenced by their environment. Environmental influences impact not just the appropriateness of the PMM elements for any given project, but also the original PMM selection process in terms of the type of PMM selected and how the PMM evolves in terms of the characteristics of the PMM (comprehensiveness). Organizations that take a holistic view of understanding the factors influencing PMM selection, evolution, and appropriateness (including the elements) for any given project are likely to see improved process success rates with the added benefit of reduced complaints about the inappropriateness of a PMM. This is more likely when the organization selects a project governance structure that is more stakeholder oriented. The findings from the fourth study show how project governance has a direct impact on both the PMM and project success.

This study's contribution to knowledge is a new comparative that allows project management to be seen in a different perspective. PMMs should be seen as dynamic sets of elements that are influenced throughout their "PMM" lives; and when understood by organizations, these elements can help them to be more effective in supporting projects and hence positively impact process success. Project governance can take a different orientation wherein the control (behavior–outcome) orientation has a direct impact on whether a PMM is comprehensive or needs supplementing and whether the corporate (shareholder–stakeholder) orientation directly impacts project success.

References

Aaltonen, K., & Sivonen, R. (2009). Response strategies to stakeholder pressures in global projects. *International Journal of Project Management,* 27(2): 131–141.

Abednego, M. P., & Ogunlana, S. O. (2006). Good project governance for proper risk allocation in public-private partnerships in Indonesia. *International Journal of Project Management,* 24(7): 622–634.

Abowd, G. D., Dey, A. K., Brown, P. J., et al. (1999). Towards a better understanding of context and context-awareness. In H.-W. Gellersen (Ed.), *Handheld and Ubiquitous Computing,* pp. 304–307. Berlin: Springer.

Adami, C., Ofria, C., & Collier, T. (2000). Evolution of biological complexity. In *Proceedings of the National Academy of Sciences,* 97, 4463–4468.

Adler, E. (1997). Seizing the middle ground: constructivism in world politics. *European Journal of International Relations,* 3(3): 319–363.

Allred, L. J., Chia, R. C., Wuensh, K. L., et al. (2007). *In-Groups, Out-Groups and Middle-Groups in China and the United States.* La Mesa, CA: National Social Science Association. Retrieved February 28, 2018, from http://www.nssa.us/journals/2007-29-1/2007-29-1-02.htm

Alvesson, M. (2009). (Post-) positivism, social constructionism, critical realism: Three reference points in the philosophy of science. In M. Alvesson & K. Sköldberg (Eds.), *Reflexive Methodology: New Vistas for Qualitative Research,* 2nd ed., pp. 15–51. London: SAGE Publications, Ltd.

Alvesson, M., & Sköldberg, K. (2009). *Reflexive Methodology: New Vistas for Qualitative Research,* 2nd ed. London: SAGE Publications, Ltd.

Alvesson, M. (1993). Organizations as rhetoric: Knowledge-intensive firms and the struggle with ambiguity. *Journal of Management Studies,* 30(6): 997–1015.

Amzaleg, Y., Azar, O. H., Ben-Zion, U., & Rosenfeld, A. (2014). CEO control, corporate performance and pay-performance sensitivity. *Journal of Economic Behavior & Organization*, 106, 166–174.

Anderson, D. K., & Merna, T. (2003). Project management strategy—project management represented as a process-based set of management domains and the consequences for project management strategy. *International Journal of Project Management*, 21(6): 387–393.

Ang, S., & Inkpen, A. (2008). Cultural intelligence and offshore outsourcing success: A framework of firm-level intercultural capability. *Decision Sciences*, 39, 337–358.

Ang, S., Van Dyne, L., & Rockstuhl, T. (2015). Cultural intelligence: Origins, conceptualization, evolution, and methodological diversity. *Handbook of Advances in Culture and Psychology*, 45, 141–164.

Ang, S., Van Dyne, L., & Tan, M. (2011). Cultural intelligence. In R. J. Sternberg and S. B. Kaufman (Eds.), *The Cambridge Handbook of Intelligence*, pp. 582–602. New York, NY: Cambridge University Press.

Ang, S., Van Dyne, L., Koh, C., et al. (2007). Cultural intelligence: Its measurement and effects on cultural judgment, decision making, cultural adaptation and task performance. *Management and Organization Review*, 3, 335–371.

Argote, L. (1982). Input uncertainty and organizational coordination in hospital emergency units. *Administrative Science Quarterly*, 27(3): 420–434.

Arnett, J. J. (2002). The psychology of globalization. *American Psychologist*, 57(10): 774–783.

Aronson, E., Wilson, T. D., & Akert, R. (2010. *Social Psychology*, 7th Ed. Upper Saddle River, NJ: Prentice Hall.

ASTD. (2012). *The Global Workplace*. Alexandria, VA: American Society for Training and Development.

Atkinson, R. (1999). Project management: Cost, time and quality, two best guesses and a phenomenon; It's time to accept other success criteria. *International Journal of Project Management*, 17(6): 337–342.

Aubry, M., Müller, R., Hobbs, B., & Blomquist, T. (2010). Project management offices in transition. *International Journal of Project Management*, 28(8): 766–778.

Avilés, L., & Purcell, J. (2012). The evolution of inbred social systems in spiders and other organisms: From short-term gains to long-term evolutionary dead ends? In H. J. Brockmann, T. J. Roper, M. Naguib, et al. (Eds.), *Advances in the Study of Behavior*, 44, 99–133. Burlington, MA: Academic Press.

Avison, D., & Fitzgerald, G. (2003). Information Systems Development: Methodologies, Techniques and Tools. EP Patent App. 20,040,768,169. 3rd ed, pp. 1–12. London: McGraw Hill.

Banik, B. J. (1993). Applying triangulation in nursing research. *Applied Nursing Research,* 6(1): 47–52.

Bar-cohen, Y. (2006). Biomimetics: Using nature as an inspiring model for human innovation. *Bioinspiration & Biomimetics,* 1(1): 1–4.

Barbaro, M. (2017, August). Germany: Wal-Mart finds that its formula doesn't work in all cultures. Berkeley, CA: CorpWatch. Retrieved March 19, 2018, from http://www.corpwatch.org/article.php?id=13969

Barron, A., & Schneckenberg, D. (2012). A theoretical framework for exploring the influence of national culture on Web 2.0 adoption in corporate contexts. *The Electronic Journal of Information Systems Evaluation,* 15(2): 76–186.

Bartunek, J. (1988). The dynamics of personal and organizational reframing. In R. E. Quinn & K. S. Cameron (Eds.), *Paradox and Transformation: Towards a Theory of Change in Organization and Management.* Cambridge, MA: Ballinger.

Basu, A. K., Lal, R., Srinivasan, V., & Staelin, R. (1985). Salesforce compensation plans: An agency theoretic perspective. *Marketing Science,* 4(4): 267–291.

Basu, S., Hwang, L.-S., Mitsudome, T., & Weintrop, J. (2007). Corporate governance, top executive compensation and firm performance in Japan. *Pacific-Basin Finance Journal,* 15(1): 56–79.

Bechara, J., & Van de Ven, A. H. (2011). Triangulating philosophies of science to understand complex organizational and managerial problems. In H. Tsoukas & R. Chia (Eds.), *Philosophy and Organization Theory,* pp. 312–342. Bingly, UK: Emerald Books.

Belassi, W., & Tukel, O. I. (1996). A new framework for determining critical success/failure factors in projects. *International Journal of Project Management,* 14(3): 141–151.

Belout, A. (1998). Effects of human resource management on project effectiveness and success: Toward a new conceptual framework. *International Journal of Project Management,* 16(1): 21–26.

Belout, A., & Gauvreau, C. (2004). Factors influencing project success: The impact of human resource management. *International Journal of Project Management,* 22(1): 1–11.

Bennis, W. G. (1969). *Organization Development: Its Nature, Origins, and Prospects. Reading,* MA: Addison-Wesley.

Benyus, J. M. (1997). *Biomimicry: Innovation Inspired by Nature. Innovation.* p. 308. New York: Harper Collins.

Bhaskar, R. (1975). *A Realist Theory of Science.* London: Taylor and Francis Group.

Biedenbach, T., & Müller, R. (2011). Paradigms in project management research: Examples from 15 years of IRNOP conferences. *International Journal of Managing Projects in Business,* 4(1): 82–104.

Biesenthal, C., & Wilden, R. (2014). Multi-level project governance: Trends and opportunities. *International Journal of Project Management,* 32(8): 1291–1309.

Blake, S. (1978). *Managing for Responsive Research and Development.* San Francisco, CA: WH Freeman & Co.

Bloch, M., Blumberg, S., & Laartz, J. (2012). Delivering large-scale IT projects on time, on budget, and on value. *Harvard Business Review.* Retrieved from http://www.mckinsey.com/insights/business_technology/delivering_large-scale_it_projects_on_time_on_budget_and_on_value Projects.pdf

Blomquist, T., Hällgren, M., Nilsson, A., & Söderholm, A. (2010). Project-as-practice: In search of project management research that matters. *Project Management Journal,* 41(1): 5–16. doi:10.1002/pmj

Boddewyn, J. (1965). The comparative approach to the study of business administration. *Academy of Management Journal,* 8(4): 261–267.

Boehm, B., & Turner, R. (2004). Balancing agility and discipline: Evaluating and integrating agile and plan-driven methods. In *Proceedings of the 26th International Conference on Software Engineering,* May 23–28, Edinburgh, Scotland.

Bond, M. H., & Hewstone, M. (1988). Social identity theory and the perception of intergroup relations in Hong Kong. *International Journal of Intercultural Relations,* 12, 153–170.

Bond, M. H. (1986). Mutual stereotypes and the facilitation of interaction across cultural lines. *International Journal of Intercultural Relations,* 10, 259–276.

Boyle, T. A., Kumar, U., & Kumar, V. (2005). Organizational contextual determinants of cross-functional NPD team support. *Team Performance Management,* 11(1/2): 27–39.

Bredin, K., & Söderlund, J. (2013). Project managers and career models: An exploratory comparative study. *International Journal of Project Management,* 31(6): 889–902.

Breese, R. (2012). Benefits realisation management: Panacea or false dawn? *International Journal of Project Management,* 30(3): 341–351.

Breitmayer, B. J., Ayres, L., & Knafl, K. A. (1993). Triangulation in qualitative research: Evaluation of completeness and confirmation purposes. *Journal of Nursing Scholarship,* 25(3): 237–243.

Brown, A., Wiele, T. Van Der, & Loughton, K. (1998). Smaller enterprises' experiences with ISO 9000. *International Journal of Quality & Reliability Management,* 15(3): 273–285.

Brown, S., & Eisenhardt, K. M. (1997). The art of continuous change: Linking complexity theory and time-paced evolution in relentlessly shifting organizations. *Administrative Science Quarterly,* 42(1): 1–34.

Bryde, D. J. (2005). Methods for managing different perspectives of project success. *British Journal of Management,* 16(2): 119–131.

Bunge, M. (1996). The seven pillars of Popper's social philosophy. *Philosophy of the Social Sciences,* 26(4): 528–556. doi:10.1177/004839319602600405

Burns, T., & Stalker, G. (1961). *The Management of Innovation.* London: Tavistock.

Burrell, G., & Morgan, G. (1979). *Sociological Paradigms and Organisational Analysis.* London: Heinemann.

Busby, J. S., & Hughes, E. J. (2004). Projects, pathogens and incubation periods. *International Journal of Project Management,* 22(5): 425–434.

Caldwell, R. (2003). Change leaders and change managers: Different or complementary? *Leadership and Organization Development Journal,* 24(5): 285–293.

Cameron, J., Sankaran, S., & Scales, J. (2015). Mixed methods use in project management research. *Project Management Journal,* 46(2): 90–104.

Cameron, K. S., & Quinn, R. E. (2006). *Diagnosing and Changing Organizational Culture.* Addison-Wesley Publishing Company, 2nd ed. San Francisco, CA: Jossy-Bass.

Cameron, K. S., Post, J. E., Preston, L. E., & Stanford, S. (2004). Book Review Essay: Effective governance in managing change: Common perspective from two lenses. *The Academy of Management Review,* 29(2): 296–301.

Cameron, R., & Molina-Azorin, J. F. (2011). The acceptance of mixed methods in business and management research. *International Journal of Organizational Analysis,* 19(3): 256–271.

Campbell, D., & Fiske, D. (1959). Convergent and discriminant validation by the multitrait-multimethod matrix. *Psychological Bulletin,* 56(2).

Campbell, D. T., Schwartz, R. D., & Sechrest, L. (1966). *Unobtrusive Measures: Nonreactive Research in the Social Sciences.* Chicago: Rand McNally.

Carlile, P., & Christensen, C. (2005). *The Cycles of Theory Building in Management Research* (No. 05-057). Boston, MA.

Castro, F. G., Barrera, Jr., M., & Martinez, Jr., C. R. (2004). The Cultural Adaptation of Prevention Interventions: Resolving Tensions Between Fidelity and Fit. *Prevention Science,* 5(1): 41–45.

Chan, A. M., & Rossiter, J. R. (2003). Measurement issues in cross cultural values research. *Proceedings of the Australia New Zealand Academy of Marketing Conference,* p. 1586. Adelaide, Australia: University of South Australia.

Cicmil, S., & Hodgson, D. (2006). New possibilities for project management theory: A critical engagement. *Project Management Journal,* 37(3): 111–122.

Clarke, A. (1999). A practical use of key success factors to improve the effectiveness of project management. *International Journal of Project Management,* 17(3): 139–145.

Clarke, T. (1998). The stakeholder corporation: A business philosophy for the information age. *Long Range Planning*, 31(2): 182–194.

Clarke, T. (2004). *Theories of Corporate Governance: The Philosophical Foundations of Corporate Governance*. London: Routledge.

Clases, C., Bachmann, R., & Wehner, T. (2003). Studying trust in virtual organizations. *International Studies of Management and Organization*, 33(3): 7–27.

Clegg, S. R. (1994). Weber and Foucault: Social theory for the study of organizations. *Organization*, 1(1).

Clegg, S. R., & Pitsis, T. S. (2002). Governmentality matters: Designing an alliance culture of inter-organizational collaboration for managing projects. *Organization Studies*, 23(3): 317–337.

Coetsee, L. (1999). *From Resistance to Commitment*. Southern Public Administration Education Foundation.

Cohen, J. (1988). *Statistical Power Analysis for the Behavioral Sciences*, 2nd ed., Vol. 2, p. 567. Hillsdale, NJ: Lawrence Erlbaum Associates.

Cooke-Davies, T. J. (2002). The "real" success factors on projects. *International Journal of Project Management*, 20(3): 185–190.

Cooke-Davies, T. J. (2004). Project management maturity models. In P. W. G. Morris & J. K. Pinto (Eds.), *The Wiley Guide to Managing Projects*. New York: John Wiley and Sons.

Cooke-Davies, T. J., & Arzymanow, A. (2003). The maturity of project management in different industries. *International Journal of Project Management*, 21(6): 471–478.

Cooper, D., & Schindler, P. (2011). *Business Research Methods*, 11th ed. Berkshire: McGraw Hill Education.

Cooper, R. G. (1999). The invisible success factors in product innovation. *Journal of Product Innovation Management*, 16(2): 115–133.

Cooper, R. G. (2007). Managing technology development projects. *IEEE Engineering Management Review*, 35(1): 67–77.

Core, J., Holthausen, R., & Larcker, D. (1999). Corporate governance, chief executive officer compensation, and firm performance. *Journal of Financial Economics*, 51(3): 371–406.

Costigan, R., Insigna, R., Berman, J., et al. (2007). A cross-cultural study of supervisory trust. *International Journal of Manpower*, 27(8): 764–787.

Crawford, L. (2006). Developing organizational project management capability: Theory and practice. *Project Management Journal*, 36(3): 74–97.

Crawford, L., & Cooke-Davies, T. J. (2008). Governance and support in the sponsoring of projects and programs. *Project Management Journal*, 39, S43–S55.

Crawford, L., & Pollack, J. (2007). How generic are project management knowledge and practice? *Project Management Journal*, 38(1): 87–97.

Crawford, L., & Cooke-Davies, T. (2012. *Best Industry Outcomes*. Newtown Square, PA: Project Management Institute, Inc.

Crawford, L., Hobbs, B., & Turner, J. (2005). *Project Categorization Systems*. Newton Square, PA: Project Management Institute.

Cronbach, L. (1951). Coefficient alpha and the internal structure of tests. *Psychometrika*, 16(3): 297–333.

Crossan, F. (2003). Research philosophy: Towards an understanding. *Nurse Researcher*, 11(1): 46–55.

Crowne, K. A. (2013). Cultural exposure, emotional intelligence, and cultural intelligence: An exploratory study. *International Journal of Cross Cultural Management*, 13(1): 5–22.

Csete, M. E., & Doyle, J. C. (2002). Reverse engineering of biological complexity. *Science*, 295(5560): 1664–1669.

Curlee, W. (2008). Modern virtual project management: The effects of a centralized and decentralized project management office. *Project Management Journal*, 39(Suppl.): S83–S96.

Danforth, B. N. (2002). Evolution of sociality in a primitively eusocial lineage of bees. *Proceedings of the National Academy of Sciences of the United States of America*, 99(1): 286–290.

Daniel, D. R. (1961). *Mangement Information Crisis. Harvard Business Review*, 39(5): 111–121.

Darwin, C. (1859). *The Origin of Species*. London: John Murray.

Davis, J. H., Schoorman, F. D., & Donaldson, L. (1997a). Davis, Schoorman, and Donaldson Reply: The distinctiveness of agency theory and stewardship theory. *Academy of Management Review*, 22(3): 611–613.

Davis, J. H., Schoorman, F. D., & Donaldson, L. (1997b). Towards a stewardship theory of management. *Academy of Management Review*, 22(1): 20–47.

Davis, M. (1971). That's interesting. *Philosophy of the Social Sciences*, 1(2): 309–355.

Dawkins, R. (1974). *The Selfish Gene*, 30th ed. Oxford, UK: Oxford University Press.

Demski, J., & Feltham, G. (1978). Economic incentives in budgetary control systems. *The Accounting Review*, 53(2): 336–359.

Dennett, D. C. (1996). *Darwin's Dangerous Idea: Evolution and the Meaning of Life*. New York: Simon & Schuster.

Denzin, N. K. (1970). *The Research Act: A Theoretical Introduction to Sociological Methods*. New York, USA: McGraw Hill.

Denzin, N. K., & Lincoln, Y. S. (2000). The discipline and practice of qualitative research. In N. K. Denzin & Y. S. Lincoln (Eds.), *The SAGE Handbook of*

Qualitative Research, 3rd ed., pp. 1–44. Thousand Oaks, CA: SAGE Publications Inc.

Diallo, A., & Thuillier, D. (2004). The success dimensions of international development projects: The perceptions of African project coordinators. *International Journal of Project Management,* 22(1): 19–31.

Dinsmore, P. C., & Rocha, L. (2012). *Enterprise Project Governance: A Guide to the Successful Management of Projects Across the Organization.* New York: AMACOM Books.

Donaldson, L. (1987). Strategy and structural adjustment to regain fit and performance: In defence of contingency theory. *Journal of Management Studies,* 24(1): 1–24.

Donaldson, L. (2001). *The Contingency Theory of Organizational Design: Challenges.* Thousand Oaks, CA: SAGE Publications Inc.

Donaldson, L. (2006). The contingency theory of organizational design: Challenges and opportunities. In *Organizational Design: The Evolving State of the Art,* 6th ed., pp. 19–40. New York: Springer Science.

Donaldson, L., & Davis, J. H. (1991). Stewardship theory or agency theory: CEO governance and shareholder returns. *Australian Journal of Management,* 16(1): 49–65.

Donaldson, T., & Preston, L. E. (1995). The stakeholder theory of the corporation: Concepts, evidence, and implications. *Academy of Management Review,* 20(1): 65–91.

Dorfman, P., Javidan, M., Hanges, P., et al. (2012). GLOBE: A twenty year journey into the intriguing world of culture and leadership. *Journal of World Business,* 47(4): 504–518.

Dougherty, D., & Hardy, C. (1996). Sustained product innovation in large, mature organizations: Overcoming innovation-to-organization problems. *Academy of Management Journal,* 39(5): 1120–1153.

Dougherty, M. J. (1998). Is the human race evolving or deevolving? Retrieved October 28, 2014, from http://www.scientificamerican.com/article/is-the-human-race-evolvin/

Dragoni, L., & McAlpine, K. 2012. Leading the business: The criticality of global leaders' cognitive complexity in setting strategic directions. *Industrial and Organizational Psychology,* 5(2): 237–240.

Dragoni, L., Oh, I., Tresluk, P., et al. (2014). Developing leaders' strategic thinking through global work experience: Moderating role of cultural distance. *Journal of Applied Psychology,* 99: 867–882.

Drouin, N., Müller, R., & Sankaran, S. (Eds.) (2013). *Novel Approaches to Organizational Project Management Research.* Copenhagen Business School Press.

Du Plessis, Y. (2011). Cultural Intelligence as Managerial Competence. *Alternation,* 18(1): 28–46.

Durward, K., II, Jeffrey, K., & Allen, C. (1998). Another look at how Toyota integrates product development. *Harvard Business Review,* 76(2): 36–47.

Earley, P. C., & Ang, S. (2003). *Cultural Intelligence: Individual Interactions Across Cultures,* Vol. 1. Stanford, CA: Stanford Business Books.

Earley, P. C., & Mosakowski, E. (2004 (October). Cultural Intelligence. *Harvard Business Review.*

Earley, P. C. (1994). Self or group? Cultural effects of training on self-efficacy and performance. *Administrative Science Quarterly,* 39: 89–117.

Easterby-Smith, M., Thorpe, R., & Jackson, P. (2008). *Management Research: Theory and Practice.* 3rd ed. London: SAGE Publications.

Easton, G. (2010). Critical realism in case study research. *Industrial Marketing Management,* 39(1): 118–128.

Eid, M., & Diener, E. (2001). Norms for experiencing emotions in different cultures: inter- and intranational differences. *Journal of Personality and Social Psychology,* 81(5): 869–885.

Eisenhardt, K. M. (1985). Control: Organizational and economic approaches. *Management Science,* 31(2): 134–149.

Eisenhardt, K. M. (1989). Building theories from case study research. *Academy of Management Review,* 14(4): 532–550.

Erez, M., & Gati, E. (2004). A dynamic, multi-level model of culture: From the micro level of the individual to the macro level of a global culture. *Applied Psychology: An International Review,* 53(4): 583–598.

Ericsson. (2013). *PROPS Manual for Project Managers.* Stockholm, Sweden: Ericsson.

Eskerod, P., & Huemann, M. (2013). Sustainable development and project stakeholder management: What standards say. *International Journal of Managing Projects in Business,* 6(1): 36–50.

Fama, E. (1980). Agency problems and the theory of the firm. *The Journal of Political Economy,* 88(2): 288–307.

Field, A. (2009). *Discovering Statistics Using SPSS,* 3rd ed. Thousand Oaks, CA: SAGE Publications.

Fiske, J. (2002). *Introduction to Communication Studies.* London, UK: Routledge, Division of Taylor & Francis.

Fitzgerald, B., Russo, N., & Stolterman, E. (2002). *Information Systems Development: Methods in Action.* Berkshire, UK: McGraw Hill Education.

Flyvbjerg, B. (2001). *Making Social Science Matter.* Cambridge, UK: Cambridge University Press.

Flyvbjerg, B., Bruzelius, N., & Rothengatter, W. (2003). *Megaprojects and Risk: An Anatomy of Ambition.* Cambridge, UK: Cambridge University Press.

Fortune, J., & White, D. (2006). Framing of project critical success factors by a systems model. *International Journal of Project Management,* 24(1): 53–65.

Fortune, J., White, D., Judgev, K., & Walker, D. (2011). Looking again at current practice in project management. *International Journal of Managing Projects in Business,* 4(4): 553–572.

Foucault, M. (1979). *Governmentality. In Discipline and Punish: The Birth of the Prison.* New York: Vintage.

Foucault, M. (1980). *Power/Knowledge: Selected Interviews and Other Writings, 1972–1977.* New York: Random House LLC.

Freckleton, R. P. (2009). The seven deadly sins of comparative analysis. *Journal of Evolutionary Biology,* 22(7): 1367–1375.

Freeman, M., & Beale, P. (1992). Measuring project success. *Project Management Journal,* 23(1): 8–17.

Friedmann, J. (1981). The active community: Towards a political-territorial framework for rural development in Asia. *Economic Development and Cultural Change,* 29(2): 235–261.

Gemünden, H. G., Salomo, S., & Krieger, A. (2005). The influence of project autonomy on project success. *International Journal of Project Management,* 23(5): 366–373.

Giebels, E., Oostinga, M. S. D., Taylor, P. J., & Curtis, J. L. (2017). The cultural dimension of uncertainty avoidance impacts police–civilian interaction. *Law and Human Behavior,* 41(1): 93–102.

Gilson, R. (1996). Corporate governance and economic efficiency: When do institutions matter. *Washington University Law Review,* 74(2).

Gittleman, J., & Luh, H. (1992). On comparing comparative methods. *Annual Review of Ecology and Systematics,* 23, 383–404.

GOA (Government Accountability Office). (2013). HUD needs to improve key project management practices for its modernization efforts, GAO-13-455, June 12, 2013). *Report to Congressional Committees.* Washington, DC.

Gompers, P., Ishii, J., & Metrick, A. (2003). Corporate governance and equity prices. *Quarterly Journal of Economics,* 118(1): 107–155.

Greene, J., & Caracelli, V. (1997). Advances in mixed-method evaluation: the challenges and benefits of integrating diverse paradigms. *New Directions for Evaluation* (Vol. 74).

Guba, E., & Lincoln, Y. S. (1994). Competing paradigms in qualitative research. In *Handbook of Qualitative Research,* 2nd ed., pp. 105–117. Thousand Oaks, CA: SAGE Publications Inc.

Guillaume, F., & Otto, S. P. (2012). Gene functional trade-offs and the evolution of pleiotropy. *Genetics,* 192(4): 1389–409.

Gupta, A. K., & Govindarajan, V. (2002). Cultivating a Global Mindset. *Academy of Management Executive,* 16(1): 116–126.

Habermas, J., & Lawrence, F. (1990). The philosophical discourse of modernity: Twelve lectures. *Contemporary Sociology,* 19(2): 316–317.

Hair, J., Black, W., Babin, B., & Anderson, R. (2010). *Multivariate Data Analysis,* 7th ed. Upper Saddle River, NJ: Prentice Hall Inc.

Hall, E. T. (1959. *The Silent Language.* Greenwich, CT: Fawcett.

Hall, E. T. (1976). *Beyond Culture.* Oxford, UK: Anchor.

Hanisch, B., & Wald, A. (2012). A bibliometric view on the use of contingency theory in project management research. *Project Management Journal,* 43(3): 4–23.

Harrington, H., Voehl, F., Zlotin, B., & Zusman, A. (2012). The directed evolution methodology: A collection of tools, software and methods for creating systemic change. *The TQM Journal,* 24(4).

Hart, O. (1995). Corporate governance: Some theory and implications. *The Economic Journal,* 105(430): 678–689.

Harvey, P., & Pagel, M. (1998). The comparative method in evolutionary biology. *Journal of Classification.* Oxford, UK: Oxford University Press.

Harzing, A. W., & Hofstede, G. (1996). Planned change in organizations: The influence of national culture. *Research in the Sociology of Organizations,* 14: 297–340.

Hastings, C. S. L., & Salkind, E. N. J. (2013). *Encyclopedia of Research Design Triangulation.* SAGE Knowledge.

Hazel, S. (2016, February 10). Why native english speakers fail to be understood—and lose out in global business. *The Conversation.* Accessed February 28, 2018, at http://theconversation.com/why-native-english-speakers-fail-to-be-understood-in-english-and-lose-out-in-global-business-54436

Hempel, C. (1965). *Aspects of Scientific Explanation and Other Essays in the Philosophy of Science.* New York/London: The Free Press.

Henrich, J., Heine, S. J., & Norenzayan, A. (2010). The weirdest people in the world? *Behavioral and Brain Sciences,* 33(2–3): 61–83.

Hernandez, M. (2012). Toward an understanding of the psychology of stewardship. *Academy of Management Review,* 37(2): 172–193.

Hirschey, M., Kose, J., & Anil, M. (Eds.). (2009). Corporate governance and firm performance. *Journal of Corporate Finance,* 6. Bingley, UK: JAI Press.

Hitt, M. (1998). Twenty-first-century organizations: Business firms, business schools, and the academy. *Academy of Management Review,* 23(2): 218–224.

Hobbs, B., Aubry, M., & Thuillier, D. (2008). The project management office as an organisational innovation. *International Journal of Project Management,* 26(5): 547–555.

Hoegl, M., & Gemünden, H. G. (2001). Teamwork quality and the success of innovative projects: A theoretical concept and empirical evidence. *Organization Science,* 12(4): 435–449.

Hofstede, G., & Bond, M. (1991). The Confucius connection: From cultural roots to economic growth. *Organizational Dynamics,* 16(4): 4–21.

Hofstede, G., & Hofstede, G. J. (2005). *Cultures and Organizations*. New York: McGraw-Hill.

Hofstede, G., Hofstede, G. J., & Minkov, M. (2010). *Cultures and Organizations: Software of the Mind*, 3rd ed. New York: McGraw-Hill Education.

Hofstede, G. (1980). *Culture's Consequences: International Differences in Work-Related Values*. Thousand Oaks, CA: SAGE Publications.

Hofstede, G. (1993). Cultural Constraints in Management Theories. *Academy of Management Executive*, 7, 81–94.

Hofstede, G. (2001). *Culture's Consequences: Comparing Values, Behaviors, Institutions and Organizations Across Nations*, p. 21. Thousand Oaks, CA: SAGE Publications.

Hofstede, G. (2011). Dimensionalizing cultures: The Hofstede Model in context. *Online Readings in Psychology and Culture*, 2(8). https://doi.org/10.9707/2307-0919.1014.

Holland, J. (1992). Complex adaptive systems. *Daedalus*, 121(1): 17–30.

Holland, J. (2012). *Signals and Boundaries: Building Blocks for Complex Adaptive Systems*. Cambridge, MA: MIT Press.

Holliday, R., & Pugh, J. (1975). DNA modification mechanisms and gene activity during development. *Science*, 187(4173): 226–232.

House, R. J., Hanges, P. J., Javidan, M., et al. (2004). *Culture, Leadership, and Organizations: The GLOBE Study of 62 Cultures*. San Francisco, CA: SAGE Publications.

Hsu, L. K. (1988. *Americans and Chinese*. Honolulu, HI: University of Hawaii Press.

Hui, C. (1988). Measurement of individualism–collectivism. *Journal of Research in Personality*, 22(1): 17–36.

Hui, C. H., & Triandis, H. C. (1986). Individualism and collectivism: A study of cross-cultural researchers. *Journal of Cross-Cultural Psychology*, 17: 225–248.

Hunt, S. D. (1990). Truth in marketing theory and research. *Journal of Marketing*, 54(3): 1–15.

Hyväri, I. (2006). Project management effectiveness in project-oriented business organizations. *International Journal of Project Management*, 24(3): 216–225.

Hyväri, I. (2006). Success of projects in different organizational conditions. *Project Management Journal*, 37(4): 31–42.

Ika, L. (2009). Project success as a topic in project management journals. *Project Management Journal*, 40(4): 6–19.

Inglehart, R., & Baker, W. E. (2000). Modernization, cultural change, and the resistance of traditional values. *American Sociological Review*, 65(2): 19–51.

Irvine, J. J., & York, D. E. (1995). Learning styles and culturally diverse students: A literature review. In J. A. Banks (Ed.), *Handbook of Research on Multicultural Education*, pp. 484–497. New York: Macmillan.

Jaenisch, R., & Bird, A. (2003). Epigenetic regulation of gene expression: How the genome integrates intrinsic and environmental signals. *Nature Genetics,* 33(Suppl.): 245–254.

Javidan, M., & Teagarden, M. B. (2011). Conceptualizing and measuring global mindset. *Advances in Global Leadership,* 6: 13–39.

Javidan, M., & Walker, J. (2013). *Developing Your Global Mindset: The Handbook for Successful Global Leaders.* Edina, MN: Beaver's Pond Press. ISBN-13: 978-1592989973.

Jensen, M., & Meckling, W. (1976). Theory of the firm: Managerial behavior, agency costs, and ownership structure. *Journal of Financial Economics,* 3(4): 305–360.

Jessen, S. A., & Andersen, E. (2000). Project evaluation scheme: A tool for evaluating project status and predicting project results. *Project Management,* 6(1): 61–67.

Jha, K. N., & Iyer, K. C. (2006). Critical determinants of project coordination. *International Journal of Project Management,* 24(4): 314–322.

Jick, T. (1979). Mixing qualitative and quantitative methods: Triangulation in action. *Administrative Science Quarterly,* 24(4).

John, K., & Senbet, L. (1998). Corporate governance and board effectiveness. *Journal of Banking & Finance,* 22(4): 371–403.

Joslin, R. (2013). The impact of new societal structures on project manager competencies and implication for project methodologies tools and techniques (PMTT). In *IPMA Expert Seminar,* Zurich, pp. 1–17.

Joslin, R. (2015). *Relationship Between Project Management Methodology (PMM), Project Success, and Project Governance.* Skema Business School.

Joslin, R., & Müller, R. (2013). A natural-science comparative to develop new insights for project management research. In N. Drouin, R. Müller, & S. Sankaran (Eds.), *Novel Approaches to Organizational Project Management Research: Translational and Transformational,* pp. 320–345. Frederiksberg, Denmark: Copenhagen Business School Press.

Joslin, R., & Müller, R. (2014). New insights into project management research: A natural sciences comparative. In *EURAM 14th Annual Conference,* pp. 1–41. Valencia, Spain.

Joslin, R., & Müller, R. (2015a). New insights into project management research: A natural sciences comparative. *Project Management Journal,* 46(2): 73–89.

Joslin, R., & Müller, R. (2015b). Relationships between project methodology and success in different governance contexts. *International Journal of Project Management,* 33(6), 1377–1392.

Joslin, R., & Müller, R. (2016a). The impact of project methodologies on project success in different contexts. In *PMI Research and Education Conference,* July 28–29, 2014, pp. 1–29. Portland, OR: Project Management Institute.

Joslin, R., & Müller, R. (2016b). The impact of project methodologies on project success in different project environments. *International Journal of Managing Projects in Business,* 9(2): 364–388. https://doi.org/10.1108/IJMPB-03-2015-0025

Joslin, R., & Müller, R. (2016c). The relationship between project governance and project success. *IJPM,* 34(4), 613–626.

Joslin, R., & Müller, R. (2016d). Identifying interesting project phenomena using philosophical and methodological triangulation. *IJPM,* 34(6), 1043–1056.

Judgev, K., & Müller, R. (2005). A retrospective look at our evolving understanding of project success. *Project Management Journal,* 36(4): 19–31.

Judgev, K., Thomas, J., & Delisle, C. L. (2001). Rethinking project management: Old truths and new insights. *International Project Management Journal,* 7(1): 36–43.

Kaiser, H. (1974). An index of factorial simplicity. *Psychometrika,* 39(1): 31.

Kaiser, L. (2006). Agency relationship and transfer pricing inefficiency. *Acta Oeconomica Pragensia,* 3, 73–81. Retrieved from http://www.vse.cz/polek/download.php?jnl=aop&pdf=94.pdf

Khan, K., Turner, J. R., & Maqsood, T. (2013). Factors that influence the success of public sector projects in Pakistan. In *Proceedings of IRNOP 2013 Conference,* June 17–19, 2013. Oslo, Norway: BI Norwegian Business School.

Khang, D., & Moe, T. (2008). Success criteria and factors for international development projects. *Project Management Journal,* 39(1): 72–84.

Kirsch, C., Chelliah, J., & Parry, W. (2011). Drivers of change: A contemporary model. *Journal of Business Strategy,* 32(2): 13–20.

Klakegg, O. J., & Haavaldsen, T. (2011). Governance of major public investment projects: In pursuit of relevance and sustainability. *International Journal of Managing Projects in Business,* doi:10.1108/17538371111096953

Klakegg, O. J., Williams, T., & Magnussen, O. M. (2009). Governance frameworks for public project development and estimation. *Project Management Journal,* 39(S1): S27–S42.

Kluckhohn, F. R., & Strodtbeck, F. L. (1961). *Variations in Value Orientation.* New York: HarperCollins.

Knorr, A., & Arndt, A. (2004, June 24). Why did Wal-Mart fail in Germany? Retrieved February 17, 2017, from http://www.iwim.uni-bremen.de/

Knorr-Cetina, K. K. D. (1981). Social and scientific method, or what do we make of the distinction between the natural and the social sciences? *Philosophy of the Social Sciences,* 11(3): 335–359.

Kong, S., & Gao, B. (2009). Chinese Executives in foreign-owned enterprises: Managing in two cultures. *Strategic Change: Briefings in Entrepreneurial Finance,* 18(3): 93–109.

Kotter, J. P., & Schlesinger, L. A. (1979). Choosing Strategies for Change. *Harvard Business Review,* 57(2): 106–114.

Kotter, J. P. (1995, May–June). Leading change: Why transformation efforts fail. *Harvard Business Review.*

Kouzes, J., & Posner, B. (1993). *Leadership Practices Inventory: A Self-Assessment and Analysis.* Expanded Edition. San Francisco, CA: Jossey-Bass.

Kramer, B. H., & Schaible, R. (2013). Life span evolution in eusocial workers— A theoretical approach to understanding the effects of extrinsic mortality in a hierarchical system. *PloS One,* 8(4): e61813.

Krieger, S. (1971). Prospects for communication policy. *Policy Sciences,* 3(2): 305–319.

Kuhn, T. (1970). The *Structure of Scientific Revolutions,* 2nd ed., Vol. 2. Chicago: The University of Chicago Press.

Kutz, M. (2017). *Contextual Intelligence: How Thinking in 3D Can Help Resolve Complexity, Uncertainty and Ambiguity.* New York, Shanghai: Palgrave MacMillan. DOI: 10.1007/978-3-319-44998-2.

Lamarck, J. B. P. A. (1838). *Histoire naturelle des animaux sans vertäbres. Histoire,* Vol. V. Paris: J. B. Bailliäre.

Langley, A., Smallman, C., Tsoukas, H., & Ven, A. H. van de. (2013). Process studies of change in organization and management: Unveiling temporality, activity, and flow. *Academy of Management Journal,* 56(1): 1–13. doi:10.5465/amj.2013.4001

Lansing, J. S. (2003). Complex adaptive systems. *Annual Review of Anthropology,* 32(1): 183–204.

Laurent, A. (1986). The cross-cultural puzzle of international human resource management. *Human Resource Management,* 25(1): 91–102.

Lawrence, P., Lorsch, J., & Garrison, J. (1967). Organization and environment: Managing differentiation and integration. *Administrative Science Quarterly,* 12(1): 1–47.

Lazonick, W., & O'Sullivan, M. (2000). Maximizing shareholder value: A new ideology for corporate governance. *Economy and Society,* 29(1): 13–35.

Lechler, T., & Geraldi, J. (2013). Comparing apples with apples: Developing a project based contingency theory. *Proceedings of IRNOP,* Oslo, June, 2013. Retrieved August 20, 2013, from http://discovery.ucl.ac.uk/1400476/

Lehtonen, P., & Martinsuo, M. (2005). Three ways to fail in project management: Role and improvement needs of a project management methodology. In *18th Scandinavian Academy of Management Meeting,* pp. 18–20. Aarhus, Denmark.

Lehtonen, P., & Martinsuo, M. (2006). Three ways to fail in project management: The role of project management methodology. *Project Perspectives,* XXVIII(1): 6–11.

Lenartowicz, T., & Johnson, J. P. (2007). Staffing managerial positions in emerging markets: A cultural perspective. *International Journal of Emerging Markets,* 2(3): 207–214.

Lewis, R. D. (2006). *When Cultures Collide: Leading Across Cultures.* London, UK: Nicholas Brealey Publishing.

Lewtas, J., Walsh, D., Williams, R., & Dobiáš, L. (1997). Air pollution exposure–DNA adduct dosimetry in humans and rodents: Evidence for non-linearity at high doses. *Mutation Research/Fundamental and Molecular Mechanisms of Mutagenesis,* 378(1–2): 51–63.

Lin, N., & Michener, C. (1972). Evolution of sociality in insects. *Quarterly Review of Biology,* 47(2): 131–159.

Linksvayer, T. (2010). Subsociality and the evolution of eusociality. In *Encyclopedia of Animal Behavior.* Academic Press.

Lipovetsky, S., Tishler, A., Dvir, D., & Shenhar, A. (1997). The relative importance of project success dimensions. *R&D Management,* 27(2): 97–106.

Maher, M., & Andersson, T. (2000). Corporate governance: Effects on firm performance and economic growth. In L. Renneboog, P. McCahery, P. Moerland, & T. Raaijmakers (Eds.), *Convergence and Diversity of Corporate Governance Regimes and Capital Markets.* Oxford, UK: Oxford University Press.

Malcom, S., & Goodship, T. H. J. (Eds.). (2001). *From Genotype to Phenotype,* 2nd ed. Oxford, UK: BIOS Scientific Publishers Ltd.

Maltz, A. A. C., Shenhar, A., & Reilly, R. R. R. (2003). Beyond the balanced scorecard: Refining the search for organizational success measures. *Long Range Planning,* 36(2): 187–204.

Markus, H. R., & Kitayama, S. (1991). Culture and self: Implications for cognition, emotion and motivation. *Psychological Review,* 98(2): 224–253.

Martins, E. P., & Garland, T. (1991). Phylogentic analyses of the correlated evolution of continuous characters–A simulation study. *Evolution,* 45(3): 534–557.

Maslow, A. H. (1970). *Motivation and Personality.* New York, NY: Harper & Row.

Matasumoto, D. (1989). Cultural influences on perception of emotion. *Journal of Cross-Cultural Psychology,* 20(1): 92–105.

Mathison, S. (1988). Why triangulate? *Educational Researcher,* 17(2), 13–17.

Maylor, H., Brady, T., Cooke-Davies, T. J., & Hodgson, D. (2006). From projectification to programmification. *International Journal of Project Management,* 24(8), 663–674.

McClelland, D. C., Atkinson, J. W., Clark, R. A., & Lowell, E. L. (1953). *The Achievement Motive.* Norwalk, CT: Appleton Century-Crofts.

McDermott, W. (2016, November). SAP's CEO on being the American head of a German multinational. *Harvard Business Review,* pp. 35–38.

McHugh, O., & Hogan, M. (2011). Investigating the rationale for adopting an internationally recognised project management methodology in Ireland: The view of the project manager. *International Journal of Project Management,* 29(5), 637–646.

McLoughlin, C. (2001). Inclusivity and alignment: Principles of pedagogy, task and assessment design for effective cross-cultural online learning. *Distance Education,* 22(1), 7–29.

McManus, J., & Wood-Harper, T. (2008). A study in project failure. Retrieved August 10, 2013, from http://www.bcs.org/content/ConWebDoc/19584

McShea, D. (1991). Complexity and evolution: What everybody knows. *Biology and Philosophy,* 6(3), 303–324.

Meissonier, R., Houzé, E., & Lapointe, L. (2014). Cultural intelligence during ERP Implementation: Insights from a Thai corporation. *International Business Research,* 7(12).

Mendel, G. (1866). Experiments in plant hybridization. *Journal of the Royal Horticultural Society of London,* (26), 1–32.

Mengel, T., Cowan-Sahadath, K., & Follert, F. (2009). The value of project management to organizations in Canada and Germany, or do values add value? Five case studies. *Project Management Journal,* 40(1), 28–41.

Merriam-Webster. (2013). Method. Retrieved November 23, 2013, from http://www.merriam webster.com/dictionary/method

Meyer, E. (2014). *The Culture Map: Breaking Through the Invisible Boundaries of Global Business.* Hachette Book Group, Imprint of Public Affairs Books. ISBN-13:9781610392501.https://www.publicaffairsbooks.com/titles/erin-meyer/the-culture-map/9781610392501/

Meyerson, D., & Martin, J. (1987). Cultural change: An Integration of three different views. *Journal of Management Studies,* 24(6), 623–647.

Miles, M., & Huberman, A. (1994). *Qualitative Data Analysis: An Expanded Sourcebook,* 2nd ed. Thousand Oaks, CA: SAGE Publications Inc.

Miles, R. E., & Snow, C. C. (1978). Organizational strategy, structure, and process. *Academy of Management,* 3(3), 546–562.

Miles, R. E., Snow, C. C., Mathews, J., & Miles, G. (1997). Organizing in the knowledge age: Anticipating the cellular form. *The Academy of Management Executive,* 11(4), 7–20.

Miller, R., & Hobbs, B. (2005). Governance regimes for large complex projects. *Project Management Journal,* 36(3), 42–50.

Milligan, B. (1986). Punctuated evolution induced by ecological change. *American Naturalist,* 522–532.

Millstein, I., & MacAvoy, P. (1998). The active board of directors and performance of the large publicly traded corporation. *Columbia Law Review,* 98(5), 1283–1322.

Milosevic, D., & Patanakul, P. (2005). Standardized project management may increase development projects success. *International Journal of Project Management,* 23(3), 181–192.

Milosevic, D., Inman, L., & Ozbay, A. (2001). Impact of project management standardization on project effectiveness. *Engineering Management Journal,* 13(4), 9–16.

Minkov, M. (2011). *Cultural Differences in a Globalizing World.* Bingley, UK: Emerald Group Publishing Limited.

Mir, F. A., & Pinnington, A. H. (2014). Exploring the value of project management: Linking project management performance and project success. *International Journal of Project Management,* 32(2), 202–217.

Mitchell, E. S. (1986). Multiple triangulation: A methodology for nursing science. *Advances in Nursing Science,* 8, 18–26.

Mitchell, M. (1998). *An Introduction to Genetic Algorithms (Complex Adaptive Systems).* Cambridge, MA: MIT Press.

Mitnick, B. M. (1973). Fiduciary rationality and public policy: The theory of agency and some consequences. In *Annual General Meeting of the American Political Science Association.* New Orleans, LA: American Political Science Association.

Mitnick, B. M. (1995). The theory of agency: The policing "paradox" and regulatory behavior. *Public Choice,* 24(1), 27–42.

Molinsky, A. (2013). *Global Dexterity: How to Adapt Your Behavior Across Cultures Without Losing Yourself in the Process.* Boston, MA: Harvard Business Review Press. ISBN-10: 1422187276 | ISBN-13: 978-1422187272.

Morgan, D. L. (2007). Paradigms lost and pragmatism regained: Methodological implications of combining qualitative and quantitative methods. *Journal of Mixed Methods Research,* 1(1), 48–76.

Morgan, G. (1997). *Images of Organization.* London: SAGE Publications.

Morris, P. W., & Hough, G. (1987). *The Anatomy of Major Projects: A Study of the Reality of Project Management.* New York: John Wiley & Sons Inc.

Morris, P. W., & Pinto, J. K. (2004). *The Wiley Guide to Managing Projects,* 1st ed. New York: John Wiley & Sons.

Morris, P. W., Crawford, L., Hodgson, D., et al. (2006). Exploring the role of formal bodies of knowledge in defining a profession—The case of project management. *International Journal of Project Management,* 24(8), 710–721.

Mustafa, G., & Lines, R. (2012). The triple role of values in culturally adapted leadership styles. *International Journal of Cross Cultural Management,* 13(1), 23–46.

Müller, R. (2009). Project governance. In R. Müller & D. Dalcher (Eds.), *Fundamentals of Project Management.* Farnham, Surrey, UK: Gower Publishing.

Müller, R., & Judgev, K. (2012). Critical success factors in projects: Pinto, Slevin, and Prescott—The elucidation of project success. *International Journal of Managing Projects in Business,* 5(4), 757–775.

Müller, R., & Lecoeuvre, L. (2014). Operationalizing governance categories of projects. *International Journal of Project Management,* 32(8), 1346–1357.

Müller, R., & Söderlund, J. (2015). Innovative approaches in project management research. *International Journal of Project Management,* 33(2), 251–253.

Müller, R., & Turner, J. R. (2007a). Matching the project manager's leadership style to project type. *International Journal of Project Management,* 25(1), 21–32.

Müller, R., & Turner, J. R. (2007b). The influence of project managers on project success criteria and project success by type of project. *European Management Journal,* 25(4), 298–309.

Müller, R., & Turner, J. R. (2010). Leadership competency profiles of successful project managers. *International Journal of Project Management,* 28(5), 437–448.

Müller, R., Andersen, E. S., Kvalnes, Ø., et al. (2013). The interrelationship of governance, trust, and ethics in temporary organizations. *Project Management Journal,* 44(4), 26–44.

Müller, R., Geraldi, J. G., & Turner, J. R. (2012). Relationships between leadership and success in different types of project complexities. *IEEE Transactions on Engineering Management,* 59(1), 77–90.

Müller, R., Pemsel, S., & Shao, J. (2014). Organizational enablers for governance and governmentality of projects: A literature review. *International Journal of Project Management,* 32(8), 1309–1320.

Müller, R., Sankaran, S., & Drouin, N. (2013). Introduction. In N. Drouin, R. Müller, & S. Sankaran (Eds.), *Novel Approaches to Organizational Project Management Research: Translational and Transformational,* pp. 19–30. Copenhagen, Denmark: Copenhagen Business School Press.

Müller, R., Turner, J. R., Anderssen, E., et al. (2014). Ethics, trust, and governance in temporary organizations. *Project Management Journal,* 45(4), 39–54.

Nachmias, D., & Greer, A. (1982). Governance dilemmas in an age of ambiguous authority. *Policy Sciences,* 14(2), 105–116.

Nelson, R. (2006). Evolutionary social science and universal Darwinism. *Journal of Evolutionary Economics,* 16(5), 491–510.

Newman, K. L., & Nollen, S. D. (1996). Culture and congruence: The fit between management practices and national culture. *Journal of International Business Studies,* 27(4), 753–779.

Ng, K. Y., Tan, M. L, & Ang, S. (2011). Global culture capital and cosmopolitan human capital. In A. Burton-Jones and J. C. Spender (Eds.), *The*

Oxford Handbook of Human Capital, pp. 96–119. Oxford, UK: Oxford University Press.

Nichols, K., Sharma, S., & Spires, R. (2011). Seven imperatives for success in IT megaprojects. Retrieved August 20, 2013, from http://www.mckinsey.com/client_service/public_sector/latest_thinking/mckinsey_on_government/seven_imperatives_for_success_in_it_megaprojects

Nisbett, R. E. (2003). *The Geography of Thought: How Asians and Westerners Think Differently . . . And Why.* New York: Free Press. ISBN-10: 0743255356 | ISBN-13: 978-0743255356.

Nowak, M., & Sigmund, K. (1993). A strategy of win-stay, lose-shift that outperforms tit-for-tat in the Prisoner's Dilemma game [letter to editor]. *Nature,* 364, 56–58. Retrieved from http://www.nature.com/nature/journal/v364/n6432/abs/364056a0.html

NSF. (2007). Enhancing support of transformative research at the National Science Foundation (NSB-07-32), p. 23. Arlington, VA. Retrieved from http://www.nsf.gov/nsb/documents/2007/tr_report.pdf

OECD. (2004). *OECD Principles of Corporate Governance 2004.* Paris: OECD Publishing.

OGC. (2002). Managing Successful Projects with PRINCE2® (2nd ed. London: The Stationery Office.

Osland, J. S., Bird, A., Delano, J., & Jacob, M. (2000). Beyond sophisticated stereotyping: cultural sensemaking in context. *Academy of Management Executive (1993–2005),* 14(1), 65–77.

Ouchi, W. (1980). Markets, bureaucracies, and clans. *Administrative Science Quarterly,* 25(1), 129–141.

Ouchi, W., & Price, R. (1978). Hierarchies, clans, and Theory Z: A new perspective on organization development. *Organizational Dynamics,* 21(4), 2–78.

OxfordDictionaries. (2014). *The Concise Oxford English Dictionary,* 12th ed. Oxford, UK: Oxford University Press.

Ozkan, N. (2007). Do corporate governance mechanisms influence CEO compensation? An empirical investigation of UK companies. *Journal of Multinational Financial Management,* 17(5), 349–364.

Packendorff, J. (1995). Inquiring into the temporary organization: New directions for project management research. *Scandinavian Journal of Management,* 11(4), 319–333.

Pastoriza, D., & Ariño, M. (2008). When agents become stewards: Introducing learning in the stewardship theory. In *1st IESE Conference on Humanizing the Firm & Management Profession,* June 30–July 2, 2008, pp. 1–16. Barcelona: IESE Business School.

Payne, J., & Turner, J. R. (1999). Company-wide project management: The planning and control of programmes of projects of different type. *International Journal of Project Management,* 17(1), 55–59.

Pearsall, J., Soanes, C., & Stevenson, A. (2011). *The Concise Oxford English Dictionary,* 12th ed. Oxford, UK: Oxford University Press.

Pearson, J. C., Lemons, D., & McGinnis, W. (2005). Modulating Hox gene functions during animal body patterning. *Nature Reviews. Genetics, 6*(12), 893–904.

Peterson, R. A. (2005). Problems in comparative research: The example of omnivorousness. *Poetics, 33*(5-6), 257–282.

Pfeffer, J., & Salancik, G. (1978). *The External Control of Organizations: A Resource Dependence Perspective.* New York: Harper & Row.

Pinto, J. K. (2014). Project management, governance, and the normalization of deviance. *International Journal of Project Management, 32*(3), 376–387.

Pinto, J. K., & Mantel, S. (1990). The causes of project failure. *IEEE Transactions on Engineering Management, 37*(4), 269–276.

Pinto, J. K., & Prescott, J. (1988). Variations in critical success factors over the stages in the project life cycle. *Journal of Management, 14*(1), 5–18.

Pinto, J. K., & Slevin, D. (1987). Critical factors in successful project implementation. *Engineering Management, 1,* 22–27.

Pinto, J. K., & Slevin, D. P. (1988). Project success: Definitions and measurement techniques. *Project Management Journal, 19*(1), 67–73.

PMI. (2013a). *Organizational Project Management Maturity Model (OPM3®),* 3rd ed. Newtown Square, PA: Project Management Institute.

PMI. (2013b). *Software Extension to the PMBOK® Guide* Fifth Edition. Newtown Square, PA: Project Management Institute.

PMI. (2013c). *The Standard for Portfolio Management,* 3rd ed. Newtown Square, PA, USA: Project Management Institute.

PMI. (2013d). *Managing Change in Organizations: A Practice Guide.* Newtown Square, PA: Project Management Institute.

PMI. (2018). *A Guide to the Project Management Body of Knowledge® (PMBOK® Guide),* 6th ed. Newtown Square, PA: Project Management Institute.

Podsakoff, P. M., & Organ, D. (1986). Self-reports in organizational research: Problems and prospects. *Journal of Management, 12*(4), 531–544.

Pogosyan, M. (2017, February 21). *Geert Hofstede: A Conversation About Culture.* Retrieved February 19, 2018, from https://www.psychologytoday.com/us/blog/between-cultures/201702/geert-hofstedeconversation-about-culture

Ralston, D. A., Egri, D. E., Stewart, S., et al. (1999). Doing business in the 21st century with the new generation of Chinese managers: A study of generational shifts in work values in China. *Journal of International Business Studies, 20*(2), 415–428.

Ralston, D. A., Gustafson, D. J., Terpstra, R. H., & Holt, D. H. (1995). Pre-post Tiananmen Square: Changing values of Chinese managers. *Asia Pacific Journal of Management, 12:* 1–20.

Ramsey, J., & Lorenz, M. P. (2016). Exploring the impact of cross-cultural education on cultural intelligence, student satisfaction and commitment. *Academy of Management Learning and Education,* 15(1), 79–99.

Renz, P. (2008). Project governance: Implementing corporate governance and business ethics in nonprofit organizations. *Journal of Management & Governance,* 13(4), 355–363.

Reuters. (2013). *Thomson Reuters Business Classification (TRBC).* New York: Reuters.

Ritter, T., & Gemünden, H. G. (2004). The impact of a company's business strategy on its technological competence, network competence and innovation success. *Journal of Business Research,* 57(5), 548–556.

Rockart, J. (1979). Chief executives define their own data needs. *Harvard Business Review,* 57(2), 81–93.

Rogers, C. P., Graham, C. R., & Mayes, C. T. (2007. Cultural competence and instructional design: Exploration research into the delivery of online instruction cross-culturally. *Educational Technology Research and Development,* 55(2), 197–217.

Sahlin-Andersson, K., & Söderholm, A. (2002). *Beyond Project Management—New Perspectives on the Temporary–Permanent Dilemma.* Copenhagen, Denmark: Copenhagen Business School Press.

Samovar, L. A., Porter, R. E., & Jain, N. C. (1981). *Understanding Intercultural Communication,* p. 24. Belmont, CA: Wadsworth.

Sanderson, J. (2012). Risk, uncertainty and governance in megaprojects: A critical discussion of alternative explanations. *International Journal of Project Management,* 30(4), 432–443.

Sauser, B. J., Reilly, R. R., & Shenhar, A. (2009). Why projects fail? How contingency theory can provide new insights: A comparative analysis of NASA's Mars Climate Orbiter loss. *International Journal of Project Management,* 27(7), 665–679.

Sayer, A. (1992). Method in social science: A realist approach. *Zhurnal Eksperimental'noi i Teoreticheskoi Fiziki.* Psychology Press.

Schein, E. H. (1992). *Organizational Culture and Leadership.* San Francisco, CA: Jossey-Bass.

Schneider, M., & Somers, M. (2006). Organizations as complex adaptive systems: Implications of complexity theory for leadership research. *The Leadership Quarterly,* 17(4), 351–365.

Schneider, S. C., & Barsoux, J. (2003. *Managing Across Cultures.* Harlow, UK: Financial Times Prentice Hall.

Schultz, R., Slevin, D. P., & Pinto, J. K. (1987). Strategy and tactics in a process model of project implementation. *Interfaces,* 17(3), 34–46.

Schultz, S. (2009). *Why Culture Matters—An Empirical Study of Working Germans and Mexicans: The Relationship Between National Culture, Resistence to Change and Communication.* Herzogenrath / Maastricht, Germany: Shaker Verlag GmbH. ISBN-10: 3832278516 | ISBN-13: 978-3832278519.

Schwartz, S. H. (2008). *Cultural Value Orientations: Nature and Implications of National Differences.* Moscow: Publ. House of SU HSE.

Scutti, S. (2015). Chinese boy, nong Yousui, can see in pitch dark: Scientists unconvinced. Retrieved from http://www.medicaldaily.com/chinese-boy-nong-yousui-can-see-pitch-dark-scientists-unconvinced-253207

Sharma, S., Durand, R., & Gur-Arie, O. (1981). Identification and analysis of moderator variables. *Journal of Marketing Research,* 18(3), 291–300.

Shenhar, A. (2001). One exploring size does not fit all projects: Classical contingency domains. *Management Science,* 47(3), 394–414.

Shenhar, A., Dvir, D., Levy, O., & Maltz, A. A. C. (2001). Project success: A multidimensional strategic concept. *Long Range Planning,* 34(6), 699–725.

Shenhar, A., Dvir, D., Milosevic, D., et al. (2005). Toward a NASA-specific project management framework. *Engineering Management Journal,* 17(4), 8–16.

Shenhar, A., Tishler, A., Dvir, D., et al. (2002). Refining the search for project success factors: A multivariate, typological approach. *R&D Management,* 32(2), 111–126.

Shenhar, A. J., & Dvir, D. (1996). Towards a typological theory of project management. *Research Policy,* 25(4), 607–632.

Shenhar, A. J., & Dvir, D. (2007). *Reinventing Project Management: The Diamond Approach to Successful Growth and Innovation.* Boston, MA: Harvard Business School Press.

Shenhar, A. J., Dvir, D., Lechler, T., & Ploi, M. (2002). One size does not fit all—True for projects, true for frameworks. In *Proceedings of PMI Research Conference,* Seattle, Washington, pp. 99–106.

Shenhar, A. J., Levy, O., & Dvir, D. (1997). Mapping the dimensions of project success. *Project Management Journal,* 28(2), 5–13.

Simon, W. (1960). Herbert Spencer and the "Social Organism." *Journal of the History of Ideas,* 21(2), 294–299.

Smith, A., & Graetz, F. M. (2011). *Philosophies of Organizational Change,* p. 32. Cheltenham, UK: Edward Elgar Publishing Inc.

Smith, H. (1975). *Strategies of Social Research: The Methodological Imagination.* Prentice Hall.

Smith, M. L. (2006). Overcoming theory-practice inconsistencies: Critical realism and information systems research. *Information and Organization,* 16(3), 191–211.

Spence, M., & Zeckhauser, R. (1971). Insurance, information, and individual action. *The American Economic Review,* 61(2), 380–391.

Stearns, F. W. (2010). One hundred years of pleiotropy: A retrospective. *Genetics,* 186(3), 767–773.

Steele, L. W. (1975). *Innovation in Big Business.* New York: Elsevier.

Steensma, H. K., Marino, L., & Dickson, P. H. (2000). The Influence of National Culture on the Formation of Technology Alliances by Entrepreneurial Firms. *Academy of Management Journal,* 43(5), 951–973.

Stevenson, T., & Barnes, F. (2001). Fourteen years of ISO 9000: Impact, criticisms, costs, and benefits. *Business Horizons,* 44(3), 45–51.

Svejvig, P., & Andersen, P. (2014). Rethinking project management: A structured literature review with a critical look at the brave new world. *International Journal of Project Management,* 33(2), 278–290.

Tannen, D. (1983). The pragmatics of cross-cultural communication. *Applied Linguistics,* 5(3), 189–195.

Tashakkori, A., & Teddlie, C. (2009). *Foundations of Mixed Methods Research. Integrating Quantitative and Qualitative.* Thousand Oaks, CA: SAGE Publications.

Tekin, A. K., & Kotaman, H. (2013). The epistemological perspectives on action research. *Journal of Educational and Social Research,* 3(1), 81–91.

The Standish Group. (2010). *Chaos Summary for 2010.* Retrieved from www.standishgroup.com

Thomas, G., & Fernández, W. (2008). Success in IT projects: A matter of definition? *International Journal of Project Management,* 26(7), 733–742.

Thomas, J., & Mullaly, M. (2007). Understanding the value of project management: First steps on an international investigation in search of value. *Project Management Journal,* 38(3), 74–89.

Thomas, M., Mitchell, M., & Joseph, R. (2002). The third dimension of ADDIE: A cultural embrace. *TechTrends,* 46(2), 40–45.

Tishler, A., Dvir, D., Shenhar, A., & Lipovetsky, S. (1996). Identifying critical success factors in defense development projects: A multivariate analysis. *Technological Forecasting and Social Change,* 51(2), 151–171.

Toivonen, A., & Toivonen, P. U. (2014). The transformative effect of top management governance choices on project team identity and relationship with the organization—An agency and stewardship approach. *International Journal of Project Management,* 32(8), 1358–1370.

Too, E. G., & Weaver, P. (2014). The management of project management: A conceptual framework for project governance. *International Journal of Project Management,* 32(8), 1382–1394.

Triandis, H. C. (1986). Collectivism vs. individualism: A reconceptualiztion of a basic concept in cross-cultural social psychology. In C. Bagley and G. K. Verma (Eds.), *Personality, Cognition and Values: Cross-Cultural Perspectives of Childhood and Adolescence.* London: Macmillan.

Triandis, H. C. (1989). The self and social behaviour in different cultural contexts. *Psychological Review,* 96, 269–289.

Triandis, H. C. (1994). Cross cultural industrial and organizational psychology. In H. C. Triandis, M. D. Dunnette, and L. M. Hough (Eds.), *Handbook of Industrial and Organizational Psychology,* pp. 103–172. Palo Alto, CA: Consulting Psychologists Press.

Triandis, H. C. (1995. *Collectivism and Individualism.* Boulder, CO: Westview Press.

Trompenaars, F., & Hampton-Turner, C. (2012). *Riding the Waves of Culture,* 3rd Ed. New York: McGraw-Hill Education. ISBN-10: 0071773088 | ISBN-13: 978-0071773089.

TSO. (2009). *Managing Successful Projects with PRINCE2®.* London: Office of Government Commerce.

Tsoukas, H., & Chia, R. (2011). Introduction: Why philosophy matters to organization theory. In M. Lounsbury (Ed.), *Philosophy and Organization Theory (Research in the Sociology of Organizations).* Vol. 32, pp. 1–21. Bingley, West Yorkshire, UK: Emerald Group Publishing Limited.

Turner, J. R. (2006). Towards a theory of project management: The nature of the project governance and project management. *International Journal of Project Management,* 24(2), 93–95.

Turner, J. R. (2007). *Handbook of Project Management.* 2nd ed. London: McGraw Hil.

Turner, J. R., Ed. (2008). The handbook of project-based management: Leading strategic change in organizations. *Annals of Physics,* 3rd ed., Vol. 54, p. 452. New York: McGraw Hill Professional.

Turner, J. R., & Cochrane, R. (1993). Goals-and-methods matrix: Coping with projects with ill defined goals and/or methods of achieving them. *International Journal of Project Management,* 11(2), 93–102.

Turner, J. R., & Keegan, A. E. (2001). Mechanisms of governance in the project-based organization: Roles of the broker and steward. *European Management Journal,* 19(3), 254–267.

Turner, J. R., & Müller, R. (2004). Communication and co-operation on projects between the project owner as principal and the project manager as agent. *European Management Journal,* 22(3), 327–336.

Turner, J. R., & Müller, R. (2006). *Choosing Appropriate Project Managers: Matching their Leadership Style to the Type of Project.* Newtown Square, PA: Project Management Institute.

Turner, J. R., Ledwith, A., & Kelly, J. (2010). Project management in small to medium-sized enterprises: Matching processes to the nature of the firm. *International Journal of Project Management,* 28(8), 744–755.

Turner, J. R., Müller, R., & Dulewicz, V. (2009). Comparing the leadership styles of functional and project managers. *International Journal of Project Management,* 2(2), 198–216.

Turner, R., Pinto, J. K., & Bredillet, C. (2011). The evolution of project management research: The evidence from the journals. In *The Oxford Handbook of Project Management,* pp. 65–106. Oxford University Press.

Van de Ven, A. H. (2007). *Engaged Scholarship*. Oxford, UK: Oxford University Press.

Vaskimo, J. (2011). Project management methodologies: An invitation for research. In *IPMA World Congress 2011,* October 12, 2011, Brisbane, Queensland. Amsterdam, The Netherlands: International Project Management Association.

Vidal, L.-A., Marle, F., & Bocquet, J.-C. (2011). Measuring project complexity using the analytic hierarchy process. *International Journal of Project Management,* 29(6), 718–727.

Vincent, J. F. V. (2001). Stealing ideas from nature—Biomimetics. In S. Pelligrino (Ed.), *Deployable Structures,* Vol. 117, pp. 51–58. Vienna: Springer-Verlag. doi:10.1038/433185a

Walker, D., Segon, M., & Rowlingson, S. (2008). Business ethics and corporate citizenship. In D. H. Walker & S. Rowlinson (Eds.), *Procurement Systems: A Cross Industry Project Management Perspective,* Vol. 27, pp. 101–139. London: Taylor & Francis.

Wateridge, J. (1998). How can IS/IT projects be measured for success? *International Journal of Project Management,* 16(1), 59–63.

Weill, P., & Ross, J. (2004). *IT Governance: How Top Performers Manage IT Decision Rights for Superior Results,* Vol. 1, pp. 63–67). Watertown, MA: Harvard Business Review Press.

Wells, H. (2012). How effective are project management methodologies: An explorative evaluation of their benefits in practice. *Project Management Journal,* 43(6), 43–58.

Wells, H. (2013). An exploratory examination into the implications of type-agnostic selection and application of project management methodologies (PMMs) for managing and delivering IT/IS projects. In *Proceedings IRNOP 2013 Conference,* June 17–19, 2013, Oslo, Norway, pp. 1–27.

Westerveld, E. (2003). The project excellence model: Linking success criteria and critical success factors. *International Journal of Project Management,* 21(6), 411–418.

Wheelwright, S., & Clark, K. (1992). *Revolutionizing New Product Development: Quantum Leaps in Speed, Efficiency, and Quality.* New York: The Free Press.

Whelan-Berry, K. S., Gordon, J. R., & Hinings, C. R. (2003). Strengthening Organizational Change Processes. *The Journal of Applied Behavioral Science,* 39(2), 186–207.

White, D., & Fortune, J. (2002). Current practice in project management—An empirical study. *International Journal of Project Management,* 20(1), 1–11.

Wibbeke, E. S., & McArthur, S. (2013). *Global Business Leadership*. London, UK: Routledge.

Wikgren, M. (2005). Critical realism as a philosophy and social theory in information science? *Journal of Documentation*, 61(1), 11–22.

Williams, M., & Vogt, P. (2011). Introduction: Innovation in social research methods. In M. Williams & P. Vogt (Eds.), *The SAGE Handbook of Innovation in Social Research Methods*, pp. 19–24. London: SAGE Publications.

Williamson, O. (1979). Transaction-cost economics: The governance of contractual relations. Journal of Law and Economics, 22(2), 233–261.

Winch, G. M. (2001). Governing the project process: A conceptual framework. *Construction Management and Economics*, 19(8), 789–798.

Wiseman, R. M., Cuevas-Rodríguez, G., & Gomez-Mejia, L. R. (2012). Towards a social theory of agency. *Journal of Management Studies*, 49(1), 202–222.

Woodward, J., Dawson, S., & Wedderburn, D. (1965). *Industrial Organization: Theory and Practice*. London: Oxford University Press.

Wright, S. (1932). The roles of mutation, inbreeding, crossbreeding and selection in evolution. In *Proceedings of the Sixth International Congress of Genetics*, Vol. 1, pp. 356–366.

Wysocki, R. K. (2011). *Effective Software Project Management*, 6th ed. Hoboken, NJ: Wiley.

Yazici, H. (2009). The role of project management maturity and organizational culture in perceived performance. *Project Management Journal*, 40(3), 14–33.

Yokoyama, S., Xing, J., Liu, Y., et al. (2014). Epistatic adaptive evolution of human color vision. *PLoS Genetics*, 10(12).

Young, P. A. (2008). The culture based model: Constructing a model of culture. *Educational Technology and Society*, 11(2), 107–118.

Yunlu, D. G., & Clapp-Smith, R. (2014). Metacognition, cultural psychological capital and motivational cultural intelligence. *Cross Cultural Management*, 21(4), 386–399.

Yusoff, W., & Alhaji, I. (2012). Insight of corporate governance theories. *Journal of Business & Management*, 1(1), 52–63.

Zoogah, D. B., & Abbey, A. (2010). Cross-cultural experience, strategic motivation and employer hiring preference: An exploratory study in an emerging economy. *International Journal of Cross Cultural Management*, 10(3), 321–343.

Zwikael, O., & Unger-Aviram, E. (2010). HRM in project groups: The effect of project duration on team development effectiveness. *International Journal of Project Management*, 28(5), 413–421.

Further Reading

Adsit, D. J., London, M., Crom, S., & Jones, D. (1997). Cross-cultural differences in upward ratings in a multinational company. *The International Journal of Human Resource Management,* 8: 385–401.

Ang, S., & Van Dyne, L. (2008). Conceptualization of cultural intelligence: definition, distinctiveness, and nomological network. *Handbook of Cultural Intelligence: Theory, Measurement, and Applications,* pp. 3–15. Boca Raton, FL: Taylor & Francis.

Ang, S., Van Dyne, L., & Tan, M. (2006). Personality correlates to the four-factor model of cultural intelligence. *Group and Organization Management,* 31: 100–123.

Berry, J. W. (1980). Acculturation as varieties of adaptation. In A. M. Padilla (Ed.), *Acculturation: Theory, Models and Findings,* pp. 9–25. Boulder, CO: Westview.

Crowne, K. A. (2008). What leads to cultural intelligence? *Business Horizons,* 51: 391–399.

Fortune Magazine. (2017). Fortune Global 500. Accessed from http://fortune.com/global500

Hmielski, K., & Ensley, M. (2007). A contextual examination of new venture performance: Entrepreneur leadership behaviour, top management team heterogeneity, and environmental dynamism. *Journal of Organizational Behavior,* 28(7): 865–889.

Huang, L., Lu, M. T., & Wong, B. K. (2003). The impact of power distance on email acceptance: Evidence from the PRC. *Journal of Computer Information Systems,* 44(1): 93–101.

Javidan, M., & Bowen, D. (2013). The global mindset of managers: What it is, why it matters, and how to develop it. *Organizational Dynamics,* 42(2): 145–155

Lazar, O. (2011). *Ensure PMO's Sustainability: Make It Temporary!* Paper Presented at PMI® Global Congress 2011, North America. Newtown Square, PA: Project Management Institute.

Lazar, O. (2016). *When Change Is Not a Change Anymore: Organizational Evolution and Improvement Through Stability.* Paper Presented at PMI® Global Congress 2016, EMEA, Barcelona, Spain. Newtown Square, PA: Project Management Institute.

Luthans, F. (2002). Positive organizational behaviour: Developing and managing psychological strengths. *Academy of Management Executive,* 16(1): 57–72.

Luthans, F., & Youssef, C. M. (2007). Emerging positive organizational behaviour. *Journal of Management,* 33(3): 321–349.

Luthans, F., Avolio, B. J., & Youssef, C. M. (2007). *Psychological Capital.* New York, NY: Oxford University Press.

Luthans, F., Avolio, B. J., Avey, J., & Norman, S. M. (2007). Psychological capital: Measurement and relationship with performance and satisfaction. *Personnel Psychology,* 60: 541–572.

Nisbett, R. E., & Masuda, T. (2003, September 16). Culture and point of view. *Proceedings of the National Academy of Sciences of the United States of America.* https://doi.org/10.1073/pnas.1934527100

Offermann, L. R., & Hellmann, P. S. (1997). Culture's consequences for leadership behavior. *Journal of Cross-Cultural Psychology,* 28(3): 342–351.

Triandis, H. C., Bontempo, R., Villareal, M., Asai, M., & Lucca, N. (1988). Individualism and collectivism: Cross-cultural perspective on self-ingroup relationships. *Journal of Personality and Social Psychology,* 54: 323–338.

Index